SITE-DIRECTED MUTAGENESIS
AND PROTEIN ENGINEERING

SITE-DIRECTED MUTAGENESIS AND PROTEIN ENGINEERING

Proceedings of the International Symposium
on Site-Directed Mutagenesis and Protein Engineering,
Tromsø, 27–30 August 1990

Editors:

M. Rafaat El-Gewely
Department of Biotechnology
Institute of Medical Biology
University of Tromsø
Tromsø, Norway

1991
ELSEVIER SCIENCE PUBLISHERS
AMSTERDAM – NEW YORK – OXFORD

© 1991 Elsevier Science Publishers B.V. (Biomedical Division)

ISBN 0 444 81431 0

This book is printed on acid-free paper.

Published by:
Elsevier Science Publishers B.V.
(Biomedical Division)
P.O. Box 211
1000 AE Amsterdam
The Netherlands

Sole distributors for the USA and Canada:
Elsevier Science Publishing Company Inc.
655 Avenue of the Americas
New York, NY 10010
USA

Library of Congress Cataloging in Publication Data:

```
International Symposium on Site-Directed Mutagenesis and Protein
   Engineering (1990 : Tromsø, Norway)
    Site-directed mutagenesis and protein engineering : proceedings of
the International Symposium on Site-Directed Mutagensis and Protein
Engineering, Tromsø, 20-30 August 1990 / editor, M. Rafaat El
-Gewely.
       p.   cm.
   Includes indexes.
   ISBN 0-444-81431-0 (alk. paper)
   1. Protein engineering--Congresses.   2. Site-specific mutagenis-
-Congresses.   I. El-Gewely, M. Rafaat.   II. Title.
   [DNLM: 1. Mutagenesis, Site-Directed--congresses.   2. Protein
Engineering--congresses.   3. Proteins--genetics--congresses.   QU 55
I672s 1990]
TP248.65.P76I58   1990
660'.65--dc20
DNLM/DLC
for Library of Congress_                                   91-9561
                                                              CIP
```

Printed in The Netherlands

INTRODUCTION

One of the biggest surprises in molecular biology is that the primary structure of proteins, as dictated by the genetic code, does not necessarily dictate a unique tertiary structure. Several proteins whose genes have been cloned and expressed by heterologous gene expression did not necessarily have the proper folding (tertiary structure). Therefore these proteins were not active. The question what are the factors controlling the formation of the active form of proteins is really the subject of our symposium.

There are no simple answers. There is no simple or unique approach to find the answer. The promise is, if and when we know the rules of protein folding, the field will be open not only to produce active proteins heterologously, but also to design new protein molecules that never existed before.

The field of protein engineering is multi-disciplinary and it is emerging as a result of interaction between fields of research such as: X-ray Crystallography, Biochemistry, Molecular Biology, and Genetics.

My biasness to Genetics for its involvement in Protein engineering is based not only on the introduction of modern techniques of molecular genetics and recombinant DNA as tools for investigation, but it is also based on the similar basic approach and philosophical goals between the two areas. The science of Genetics was established on the basis of studies dealing with genes through the study of their variation, an important aspect of protein engineering, namely structure-function relationship, is based on the study of a given protein through the study of its variation. 'Variations' in genetics research or in protein engineering could be naturally occurring mutants affecting the phenotype/protein function, or they could be introduced. If the science of Genetics evolved rapidly allowing us to be able to do various genetic manipulations and control, the hope is that protein engineers of the future will be able to design proteins of any needed function.

M. Raafat El-Gewely
Tromsø, Norway

ACKNOWLEDGEMENTS

This Symposium would not have been possible without the generous support of *PHARMACIA BIOSYSTEMS A/S*, and *The Erna and Olav Aakre Foundation for Cancer Research*.

We would also like to acknowledge the support of the following organizations:

The University of Tromsø
The Norwegian Cancer Society
International Union of Biochemistry (IUB)
The Norwegian Society for Science and Humanities (NAVF)
Perkin-Elmer Cetus Analytical Instruments, Germany
The City of Tromsø

A great number of people helped in organizing this symposium. We would like to extend our thanks to Tove Eriksen Heen, Nils Peder Willassen, Norunn Storbakk and to Inge Nilsen for this remarkable help in all the relevant details.

CONTENTS

viii

MOLECULAR STRUCTURE AND DYNAMICS

© 1991 Elsevier Science Publishers B.V. (Biomedical Division)
Site-Directed Mutagenesis and Protein Engineering
M.R. El-Gewely, editor.

MOLECULAR STRUCTURE AND DYNAMICS
INTRODUCTION

ASBJØRN HORDVIK

Protein Crystallography Group, IMR, University of Tromsø, N-9000 Tromsø, (Norway)

Why are we trying to determine the detailed three-dimensional structure of important biological macromolecules, - like enzymes for example? - Simply because we are convinced that in order to really understand their function and behaviour it is imperative to be able to know them and study them at molecular level.[1,2]

Once the 3D structure of a macromolecule is known in sufficient detail, one may be able to change it a little bit, chemically or by site-directed mutagenesis, to taylor its function for specific purposes.

Such work may, when combined with Molecular Dynamics[3] simulations, reveal valuable information about the interaction between enzyme and substrate for example, and by the latter method one can also seek possible solutions to drug binding problems.

Although one realizes that it is possible to determine the 3D structure of proteins by NMR methods, it should be underlined that the method which so far is most widely used, and far superior with respect to accuracy, is X-ray crystallography[4,5].

Reports from structure work on biological macromolecules will now be presented, and it is encouraging to see how structure chemistry and biology have met on the molecular level in order to disclose the secrets of biological mechanisms.

REFERENCES
1. Rees, A. R., Sternberg, M. J. E. (eds) From Cells to Atoms. Blackwell Scientific Publications, Oxford, (1987).

2. The Molecules of Life. Readings from Scientific American, W. H. Freeman and Company, New York, (1986).

3. Karplus, M., Petsko, G. A. (1990) Nature 347:631-639.

4. Stryer, L. (ed) Biochemistry. W. H. Freeman and Company, New York, (1988), pp 59-62.

5. Richardson, J. S. (1981) In: Anfinsen, C. B., Edsall, J. T., Richards, F. M. (eds), Advances in Protein Chemistry. Academic Press, New York, pp 167-339.

© 1991 Elsevier Science Publishers B.V. (Biomedical Division)
Site-Directed Mutagenesis and Protein Engineering
M.R. El-Gewely, editor.

MOLECULAR STRUCTURE OF NEUROTRANSMITTER RECEPTORS AND LIGANDS

SVEIN G. DAHL

Department of Pharmacology, Institute of Medical Biology,
University of Tromsø, N-9001 Tromsø, Norway

MOLECULAR MODELLING

Computer graphics and molecular modelling techniques have
undergone a rapid development over the last 5 years, and many
laboratories involved in molecular biology research or drug
design have recently acquired computer graphics equipment with
software for molecular modelling. This is reflected in the
increasing number of reports on a wide range of applications of
molecular modelling in medicinal chemistry and biology, which has
lead to development of specific guidelines for such publications
(1). It has been anticipated that worldwide expenditure in
computational chemistry will grow from a 1987 level of $125
million to $770 million in 1992, a large proportion of which will
be funded by the drug industry (2).

Applications of molecular modelling techniques have become
feasible by the recent development in computer technology, which
has provided high-performance workstations with increasing power
at steadily lower prices. Modern workstations offer high quality
raster or vector colour graphics, perspective, depth cueing,
three dimensional clipping and real-time translation and
rotation. Combined with computational chemistry methods, these
techniques have a large potential for providing insight into the
spatial arrangements of atoms in molecules, the charge distribution over molecules, and the dynamics of molecular interactions. Molecular modelling methods may thus provide important
information about the molecular mechanisms of action of drugs and
other biologically active compounds.

Molecular Mechanics and Molecular Dynamics Calculations

X-ray diffraction techniques have been and still are the
most widely used experimental methods for determination of three
dimensional molecular structures. Many computational methods used
in molecular modelling are based on atomic coordinates from
crystal structures, as indicated in Fig. 1.

In our recent studies on the molecular conformations and
dynamics of psychotropic drug and neurotransmitter molecules, we

have used the Molecular Interactive Display And Simulation
(MIDAS) programs (3) for molecular graphics, with an Evans and
Sutherland PS390 workstation and a DEC Microvax II/Ultrix system
as the host machine. Water-accessible molecular surfaces and
electrostatic potentials 1.4 Å outside the surfaces were calculated with the MIDAS programs, and the potentials were illustrated
by colour coding of the surfaces as shown in Fig. 2.

Molecular mechanical geometry optimization and molecular
dynamics simulations were performed with the AMBER 3.0 programs
(4,5), using the all atom force field. Molecular mechanical
calculations provide energy minimized atomic coordinates from a
starting structure. Several different force fields are available,
and the resulting structure may somewhat depend on the force
field and the parameters used for bond lengths, angles, torsional
barriers and non-bonded interactions between atoms.

The internal movements of biologically active molecules in
solution occur on a picosecond time scale or even faster.
Molecular dynamics simulations, which are used to investigate the
trajectories of molecular movements across conformational
barriers, have provided new insight into the molecular movements
and functioning of neurotransmitter receptor ligands (6-12).
Combined with modern computer graphics techniques, such simulations provide a new view of the mechanisms of molecular interactions, which may be more valid than static geometric concepts
derived from crystal structures and potential molecular energy
calculations alone.

As indicated in Fig. 1, crystal structures of drugs and
neurotransmitter molecules were initially refined by molecular
mechanical energy minimization, and the refined structures were
used as starting coordinates for molecular dynamics simulations.
For some compounds where no crystal structure was available, an
initial model was constructed from the structure of a similar
molecule, using the computer graphics system and the MIDAS
programs, as indicated in the right part of Fig. 1. The new model
was then refined by energy minimization. This strategy was used
to construct models of *trans*(E)-chlorprothixene (9), amitrip-
tyline (11) and nortriptyline (11), for which no crystal structures have been reported.

Fig. 1. Molecular modelling procedure. Crystal structures were
initially refined by molecular mechanical energy minimization
(EMIN), based on the AMBER force field (4,5). Energy minimized
structures were used as starting points for molecular dynamics
simulations (MDYN), based on the AMBER force field. The MIDAS
programs (3) were used for molecular graphics.

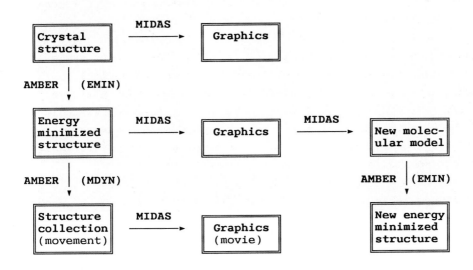

MOLECULAR STRUCTURE AND DYNAMICS OF RECEPTOR LIGANDS

The molecular modelling techniques described above were
used to study the three dimensional structures, low-energy
conformations and molecular dynamics of the neurotransmitters
acetylcholine (6), serotonin (7) and dopamine (8), and a series
of tricyclic drugs including neuroleptics of the *cis*(Z)-thioxan-
thene type and their inactive *trans*(E)-isomers (9,10), various
tricyclic antidepressants (11), and the four isomers of a
metabolite of one of these, 10-hydroxy nortriptyline (12).

As shown by an example in Fig. 2, computer graphics
techniques were used to illustrate three dimensional molecular
structures and the distribution of positive and negative
molecular electrostatic potentials over molecular surfaces. The
electrostatic potentials surrounding the dopamine molecule in the
energy minimized *anti* conformation, were strongly positive around
the terminal part of the side chain containing a protonated
dimethylamino group (Fig. 2). Neutral and slightly positive
electrostatic potentials surrounded most of the aromatic ring,

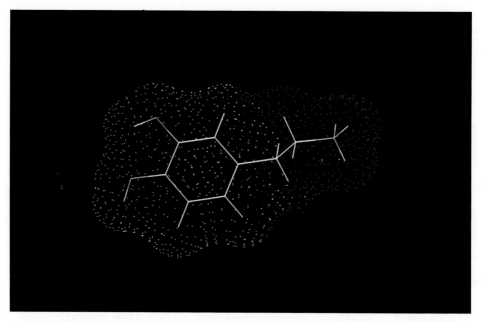

Fig. 2. Energy minimized *anti* conformation of dopamine with water accessible molecular surface, coloured according to molecular electrostatic potentials (e, kcal/mol) 1.4 Å outside the surface: Red e > 15, white 7 ≤ e ≤ 15, blue 0 ≤ e < 7, yellow e < 0. Colour coding of atoms: nitrogen - green, oxygen - blue, hydrogen and carbon - white.

and a small area near the oxygen atom of the m-hydroxyl group had negative electrostatic potentials as low as -5 kcal/mol (8).

Specific binding of ligands to macromolecules usually involves enclosure of most of the ligand molecule by a binding pocket. Electrostatic interactions may contribute to the stabilization of such complexes, and molecular electrostatic potentials therefore provide useful information about the mechanisms of ligand-receptor interactions.

Various side chain conformations of the drug and neurotransmitter molecules were observed during the molecular dynamics simulations, which were carried out *in vacuo* and in aqueous solution. The motions between different side chain conformations followed quite unexpected trajectories, with twisting of the whole molecule such that most of its mass was kept in place, rather than rotating around single bonds. The tricyclic ring systems showed unexpectedly high flexibility, with variations between 100° and 180° in the angle between the phenyl rings.

TABLE I. Neurotransmitter receptors with primary structure known from cloning experiments (19,20)

Superfamily	
Ion channel gating	G protein coupled
$GABA_A$	Dopamine D_1, D_2, D_3
Serotonin $5\text{-}HT_3$	Serotonin $5\text{-}HT_{1A}$, $5\text{-}HT_{1C}$, $5\text{-}HT_{1D}$, $5\text{-}HT_2$
Nicotinic acetylcholine	M_1, M_2, M_3, M_4, M_5 Muscarinic acetylchol.
	$\alpha_{1A}\text{-}$, $\alpha_{1B}\text{-}$, $\alpha_{2A}\text{-}$, $\alpha_{2B}\text{-}$ Adrenergic
	$\beta_1\text{-}$, $\beta_2\text{-}$, $\beta_3\text{-}$ Adrenergic

These studies have provided new insight into the molecular structure and dynamics of compounds acting as agonists or antagonists on different neurotransmitter receptors.

STRUCTURE OF NEUROTRANSMITTER RECEPTORS

A series of neurotransmitter receptors have been cloned and their amino acid sequence deduced from the DNA sequence. This way modern molecular biology has further clarified and confirmed the subtypes within various families of neuroreceptors, which previously were classified based on the potencies and affinities of various agonists and antagonists. The cloning and sequencing of neurotransmitter receptors has also demonstrated that they may be divided into different superfamilies, as indicated in Table I. Furthermore, that several segments of their peptide chain traverse the cell membrane, connected by intra- and extracellular loops of different lengths.

However, so far there are no available experimental data on the detailed three dimensional structure of any neurotransmitter receptor molecule. The only membrane protein for which a detailed three dimensional crystal structure has been reported, is the photosynthetic reaction center of *Rhodopseudomonas viridis*, which has three sub-units containing a total of 11 membrane spanning alpha helices (13).

The receptors of the superfamily which transfer signals into cells via guanine nucleotide binding regulatory proteins (G proteins), have several common structural features, including seven membrane spanning helices. There is relatively high sequence homology in the membrane spanning domains between the different families of G protein coupled receptors. Site-directed mutagenesis experiments have demonstrated that two aspartate residues in transmembrane segments 2 and 3 represent binding sites for agonists and antagonists, respectively, in the β_2-adrenergic (14) and M_1 muscarinic acetylcholine (15) receptors. These residues are conserved in all the G protein coupled receptors listed in Table I, including the dopamine D_1, D_2 and D_3 receptors.

The identification of ligand binding sites near the middle of the second and third membrane spanning domains, has lead to the hypotheses that the β_2-adrenergic (16) and M_1 muscarinic acetylcholine receptors (15) form a hydrophilic core by an circular arrangement of the seven transmembrane helices, similar to that observed in the purple membrane of bacteriorhodopsin (17).

We have postulated a similar arrangement of the seven transmembrane helices in the dopamine D_2 receptor, which is assumed to be the primary target of action of antipsychotic drugs. A receptor model was constructed from the amino acid sequence of the rat dopamine D_2 receptor (18), and refined by molecular mechanical energy minimization and molecular dynamics simulations. Its electrostatic potentials suggest that the primary interaction of cationic agonist and antagonist ligands with the receptor is electrostatic, with negatively charged aspartate residues at the synaptic side and in helix 2 and 3.

The virtually inactive *trans*(E)-thioxanthenes have strong negative electrostatic potentials around the 2-substituent on the ring system, contrary to the pharmacologically active *cis*(Z)-isomers (9,10). It seems likely, therefore, that the negative molecular electrostatic potentials around a part of the *trans*(E)-thioxanthene molecules may weaken their initial electrostatic interactions with the dopamine D_2 receptor. This provides a possible explanation of why *trans*(E)-isomers of thioxanthenes are so much less active than *cis*(Z)-isomers in dopamine receptor binding and related tests.

As shown by this example, molecular modelling techniques may extend our knowledge on the molecular function of receptor proteins, even in the absence of detailed x-ray crystallographic or other experimental data.

REFERENCES

1. Cohen NC, Blaney JM, Humblet C, Gund P, Barry DC (1990) J Med Chem 33:883-894

2. Glen RC (1990) Drug News Perspect 3:332-336

3. Ferrin TE, Huang CC, Jarvis LE, Langridge R (1988) J Mol Graphics 6:2-12; *ibid* 13-27

4. Weiner SJ, Kollman PA, Nguyen DT, Case DA (1986) J Comp Chem 7:230-252

5. Singh UC, Weiner PK, Caldwell JW, Kollman PA (1986) Assisted model building with energy refinement. AMBER UCSF Version 3.0. Dept. Pharmaceutical Chemistry, University of California, San Francisco, CA 94143

6. Edvardsen Ø, Dahl SG (1990) J Neural Transm, *in press.*

7. Edvardsen Ø, Dahl SG (1990) Mol Brain Res, *in press.*

8. Edvardsen Ø, Dahl SG (1991) Submitted for publication.

9. Sylte I, Dahl SG (1991) J Pharm Sci, *in press.*

10. Sylte I, Dahl SG (1991) Pharm Res, *in press.*

11. Heimstad E, Edvardsen Ø, Ferrin TE, Dahl SG (1991) Submitted for publication.

12. Heimstad E, Edvardsen Ø, Dahl SG (1991) Neuropsycho-pharmacology, *in press.*

13. Deisenhofer J, Epp O, Miki K, Huber R, Michel H (1985) Nature 318:618-624

14 Strader CD, Sigal IS, Register RB, Candelore MR, Rands E, Dixon RAF (1987) Proc Natl Acad Sci USA 84:4384-4388

15. Fraser CM, Wang C-D, Robinson DA, Gocayne JD, Venter JC (1989) Mol Pharmacol 36:840-847

16. Dixon RAF, Sigal IS, Rands E, Register RB, Candelore MR, Blake AD, Strader CD (1987) Nature 326:73-77

17. Henderson R, Unwin PNT (1975) Nature 257:28-32

18. Bunzow JR, Van Tol HHM, Grandy DK, Albert P, Salon J, Christie M, Machida CA, Neve KA, Civelli O (1988) Nature 336:783-787

19. Strosberg AD (1991) Eur J Biochem (Review), *in press.*

20. Zgombic JM, Weinshank RL, Macchi M, Hartig P, Branchek TA (1990) Soc Neuroscience Abstr 16:1195.

© 1991 Elsevier Science Publishers B.V. (Biomedical Division)
Site-Directed Mutagenesis and Protein Engineering
M.R. El-Gewely, editor.

The Active Site and Catalytic Mechanism in Phospholipase C from *Bacillus cereus*

Edward Hough

Institute of Mathematical & Physical Sciences, University of Tromso, N9000 Tromso, Norway.

Introduction

Phospholipase C (PLC) from *Bacillus Cereus* is a monomeric exocellular Zn-metallo-enzyme[1]. The PLC gene has been cloned and sequenced showing that the enzyme is synthesised as a 283 residue precursor with a 24 residue signal peptide and a 14 residue propeptide[2]. The secreted form of PLC contains 245 amino acids with $M_r = 28520D$. Since synthesis of PLC, Alkaline Phosphatase and a sphingomyelinase-C by *B. cereus* is repressed by P_i the enzyme may be part of a bacterial phosphate retrieval system[3]. PLC is highly stable, even in the presence of 8M urea at 40^oC[4]. Stability is closely linked to the presence of Zn^{2+} and the enzyme may be reversibly unfolded/refolded in guanidium hydrochloride by removal/ addition of this metal ion[5]. The Zn^{2+} ions may be exchanged with other metals, common feature of metalloenzymes where the metal ions are involved in catalysis, eg Alkaline Phosphatase[6].

The enzyme shows a marked preference for phosphatidylcholine although it is active against phosphatidylethanolamine and phosphatidylserine[7]. PLC is inactive against cardiolipin and lysolipids but shows a low level of activity against sphingomyelin[8] which is increased by replacement of one Zn^{2+} ion by Co^{2+} [9]. Plots of activity against substrate concentration are characteristic of enzymes where activity is influenced by the aggregation state of the substrate, with an approximately 30-fold increase in Kcat/Km as the substrate concentration passes the critical micelle concentration[10, 11]. Since it cleaves the polar head groups from membrane phospholipds the enzyme is widely used as a tool for the study of cell membranes. In recent years a wide spectrum of mammalian PLC have been reported including a Ca^{2+} dependent neutral phosphatidylcholine (PC) specific PLC of M_r 29000D in canine myocardial cytosol[12], a Ca^{2+} independent PC-PLC in bovine pulmonary endothelial cells[13] and in a wide variety of tissues from the rat[14]. A PC-PLC has also been isolated from a human monocytic cell line using antibodies raised against *B. cereus* PLC[15] suggesting that there may be a similarity in structure between mammalian and bacterial PLC's. The involvement of PC-PLC's in the generation of second messengers has been proposed[16] and PLC has also been shown to stimulate enhanced prostaglandin biosynthesis to an extent which is similar to Phospholipase A_2[17]. PLC from *B. cereus* may thus represent a useful model for the hitherto poorly characterised mammalian PLC´s.

The Structure of PLC

The crystal structure of PLC has been solved at 1.5Å resolution[18] and has been refined to R=15% including 243 water molecules. Some impression of the quality of the structure determination may be gained from the portion ofelectron density which is shown in plate 1 where among other details one can clearly see the unambiguous envelope conformation of the pyrrole ring in Pro 59.

14

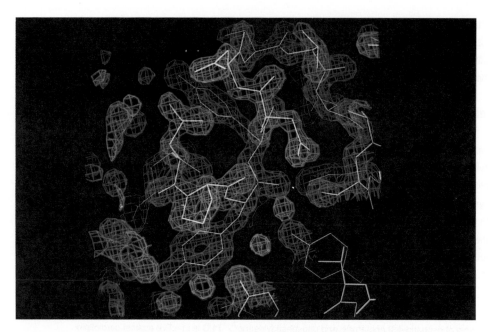

Plate 1. Electron density for residues 57 to 67. The small spheres of density are due to bound water molecules

Plate 2. Surface representation of the PLC molecule. The metal ions are shown a green spheres and the fold of the protein chain as a ribbon.

PLC is an all-helix enzyme and belongs to a novel structural class (see figure1). Helices A, B, C, D, F & H_2 form a compact arrangement of mixed parallel- and anti-parallel helices with a predominantly right-handed twist. Helix E lies across the loop between helices F & G and the bend between helices H_1 & H_2.

Identification of the Active Site

Determination of the crystal structure of a protein leads to a 3-dimensional image of the molecule, but does **not** always lead to the direct identification of the catalyic or substrate binding site in the enzyme (e.g. the newly reported structure of two triacylglycerol lipases[19, 20] where the active sites are concealed). Location of the active or binding site proceeds by one or several of the following routes :

a)   Location of a cleft in the molecular surface.

A surface representation of the PLC molecule is shown in plate 2. With the exception of a single 5Å wide by 8Å deep cleft between helix E and the ends of helices A, B2, C and D the surface of the PLC molecule is smooth. It seems likely that this cleft forms the active site of the enzyme.

b)   Location of metal ions known to be involved in catalysis.

The three metal ions in PLC lie on the inner surface of the cleft. Since PLC is deactivated by the removal of one metal ion (the "catalytic" Zn) and shows changes in substrate specficity when Zn^{2+} is replaced by other metal ions[21] it seems likely that this cleft is the active site.

The metal ions form a cluster (see Figure 2) which is remarkably similar to that in Alkaline Phosphatase (AP) from *E. coli*.[6] Asp122 forms a carboxylato-bridge between Zn1 and Zn3 with Zn-Zn distance 3.3Å. This bridge appears to be similar to that formed by Asp51 in Alkaline phosphatase and possibly by Asp355 in the Klenow fragment[22]. An OH⁻ ion forms a second bridge between Zn1 and Zn2   The absence of such a second bridging ligand in Klenow fragment permits the "more comfortable" metal-metal distance of 4.3Å. Zn3 is also coordinated to the amino- and carboxyl-groups of the N-terminal Trp to form a five membered chelate ring similar to that observed in avian pancreatic polypeptide[23] and in several Zn-amino acid complexes eg. with glutamic acid[24]. All three Zn(II) ions in PLC are 5-coordinate and have approximately trigonal bipyramidal geometry.

The Zn^{2+} ions crosslink widely disparate parts of the protein chain which probably explains the high conformational stability of the holoenzyme relative to the apo-enzyme[25].

c)   Location of surface region with appropriate hydrophobic or hydrophilic properties.

Figure 3 shows the active site cleft. The hydrophobic and neutral residues Ser 64, Thr65, Phe66, Ile 80, Thr133, Asn134 and Ser143 which form the inner surface of the cleft are emphasised. The figure also shows
the location of Tyr52, the disordered Tyr 56/356, Tyr61 and Tyr79. With the exception of Glu4 and single carboxyl oxygens from Asp55 and Glu146 all the polar side chains in the cleft are liganded to the metal ions

16

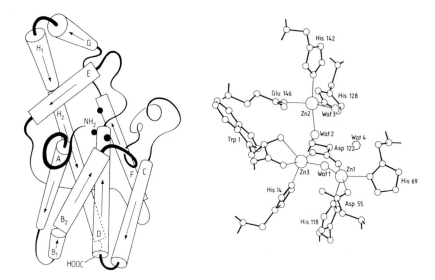

Figure 1. The PLC fold. α-helices are shown as cylinders and metal ions as filled circles.

Figure 2. Metal coordination in PLC including coordinated water molecules.

and are largely inaccessible to substrate. The 10 water molecules found in the active site lie in the region around the metal ions and their ligands implying that the remainder of the active site is indeed hydrophobic. An intermolecular contact region involving the outer parts of tthe B-C and D-E loops is also mediated by hydrophobic residues (Ile80, Pro81, Phe82, Val 135 & Thr133).
Although less marked, a similar differentation into polar and non polar regions is present in the active site of the PLA2[26]. It is interesting to note that all 9 ligands and most of these active site residue are conserved in the α-toxin (a PC-PLC) from *Clostridium perfringens* [27], [28], [29].

d)   Structure determination of PLC/inhibitor complexes

The identification of the active site is supported by results from crystallographic studies of PLC which has been soaked with the inhibitors :

1) potassium iodide - monovalent anions inhibit PLC[30]. An iodide ion was found in the cleft 5Å from Zn1 and 7.5 Å from Zn2[31], There is no obvious counter charge for this ion.

2) The metal chelating buffer compound **"Tris"** (tris-hydroxymethyl-aminomethane), also an inhibitor[30]. Our 1.9A study has shown that a 'tris' molecule lies in the cleft with an N-Zn2 distance of 2.4Å, an N-Wat1 (fig.2)

3) inorganic phosphate - again an inhibitor. Our 1.9Å study has shown that a clearly defined phosphate group is located close to all three metal ions with phosphate oxygens replacing Wat1 and Wat2 (see fig. 2) and two other water molecules in the cleft.

Substrate Binding

The metal cluster in the active site of PLC carries a net positive charge which will clearly be attractive

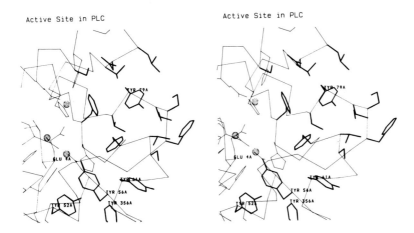

Figure 3. Stereo drawing emphasising the hydrophobic nature of much of the active site in PLC.

to the phosphate group in a substrate molecule. If one **assumes that a lipid phosphate group binds to the metal ions in a similar way to that found in our study of P$_i$PLC,** there are clearly two possible orientations for the substrate molecule. These can be seen in figure 4 where R$_1$ and R$_2$ represent in one orientation the diacylglycerol (DAG) and phosphorylcholine (PC) groups, respectively, and *vice versa* for the second orientation. The major difference between these orientations lies in the amphiphilic nature of the active site and raises questions about the relative hydrophobicity of the DAG and PC moieties in a phospholipid molecule.

The crystal structures of a number of phospholipids have been solved by X-ray crystallographic methods and are reviewed by Hauser et al.[32.] This includes several of the natural substrates for PLC and has enabled a distance of 2.7AÅand with two of its hydroxyl-groups forming hydrogen bonds to the backbone N-H of Asn134 via a water molecule. simple molecular modelling study based on the assumption that the phosphate group in the substrate binds to the metal ion cluster in PLC in a similar way to that found in our study of P$_i$-PLC. A similar proposal has been made for Alkaline phosphatase where the phosphomonoester substrate has been modelled into electron density[6]. The same applies to difference electron density observed in crystals of Klenow fragment which had been treated with deoxycytidine monophosphate and deoxythymidine tetra-nucleotide[22] where a substrate phosphate group is coordinated to both (see above) metal ions. The high affinity of phosphate for Zn^{2+} is also apparent in the binding of the inhibitor phosphoramidon to Zn^{2+} in the active site of thermolysin[33] with a phosphate oxygen replacing a coordinated water molecule.

In orientation 1 the DAG-moiety lies in the polar region displacing at least 6 water molecules and offering the possibility of protein to ester carbonyl-group hydrogen-bonding. The Choline-group lies in the hydrophobic region in a similar position to that found for Tris and the I$^-$ ion (see above). Although the Choline-group is positively charged it is predominantly aliphatic in nature and has a diffuse charge due to the inductive effect of its three methyl groups. Tyr79 does offer the possibility of a balancing negative charge although it is rather remote (5A) in the native structure. In

18

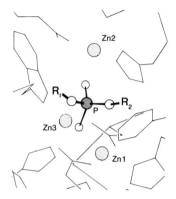

Figure 4. Proposed docking of phospholipid in the active site of PLC.

the phosphorylcholine binding Immunoglobulin Fab McPC603 complex reported by Satow et al.[34] the ligand lies in an environment which is similar to that for orientation 1. There are no serious collisions between the enzyme and substrate in this orientation and the lipid is only minimally distorted from its conformations in single crystals (Hauser *et al.*,1980).

In orientation 2 the Ch-group is adjacent to Glu4 and Asp150 and the DAG-moeity lies tightly in the hydrophobic cleft with few possibilities for hydrogen-bonding. Docking of substrate in this orientation requires the movement of several active site side chains and a more strained lipid conformation.

In both orientations the phosphorylcholine group and the ester linkages are buried in the active site cleft but only the first 1-2 methylene groups in the fatty acid side chains are in contact with the protein. This agrees well with the results of **structure/activity** studies (see below).

Binding of substrate via the phosphate-metal complexation described above will result in considerable delocalisation of electrons away from the phosphorus atom rendering it highly susceptible to nucleophilic attack by free water or by water which has been activated by an electron-donating protein side chain. However, details of this process are not clear at present.

How to find out what happens

Among methods which are available :

a) **Structure/activity studies** using a range of different substrates - in studies of the effect of structural modificationsof PC's on their suitability as substrates for PLC, Roberts and her coworkers[35, 36, 11] have shown that a) "the fatty acid side chains must be long enough to contain a hydrophobic region which can bind to the enzyme", b) replacement of the ester linkages by ether linkages results in dramatically reduced activity thus implicating the ester carbonyl groups in substrate binding and c) that the introduction of bulky side chains on the first two methylene groups in the fatty acids influences activity. Substitution further along the fatty acid chains has no effect. In a related study Snyder et al.[37] have confirmed that the presence of an ester group is essential for substrate binding and that the enzyme is selective for the R absolute conformation at C2 in the glycerol moiety. Our simple molecular modelling studies (see above) agree well with these conclusions.

b) Treatment of crystals with inhibitors -- see above.

c) Treatment of crystals with reaction products

d) Trapping substrate in the active site of the crystalline enzyme by using :

>1) Poor substrates

>2) Low temperature

>3) Very rapid data collection methods

e) Treatment of crystals with substrate analogues.

The application of methods c - e to PLC will be discussed below.

Crystallographic Experiments to Determine Substrate Orientation
PLC catalyses the following reaction :

Phosphatidyl-X $+ H_2O \longrightarrow$ 1,2-Diacylglycerol + Phosphoryl-X
where X is choline, serine or ethanolamine.
In simple terms the reaction pathway can be presented thus :

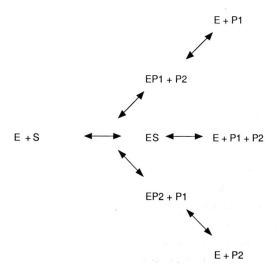

where S = phospholipid or substrate analogue, P1 = phosphoryl-choline, -serine or ethanolamine and P2 = diacylglycerol :

Treatment of crystals of PLC (or co-crystallisation of PLC) with S, P1 or P2 could yield crystalline complexes whose solution would solve the question of substrate orientation in the active site of PLC and almost certainly lead to the elucidation of the catalytic mechanism. However this approach is limited by the following factors :

a)   **Availablity of suitable ligands.**

b)   **Limits imposed by crystal packing and stability** i.e.

1)   Diffusion into the crystal - the "channels" large enough?

2)   Is the active/binding site accessible in the crystal?

3)   Solubility of the ligand in high salt concentrations?

4)   Mechanical stability of the crystals after treatment with a ligand.

5)   Stability under X-ray data collection - often much poorer after treatment with a ligand.

6)   Stability of the ligand when bound to the crystalline enzyme.

A fortunate aspect of crystalline PLC is that the active site of the enzyme is unhindered by neighbouring molecules and that channels through the crystals are large enough to permit perfusion with quite large ligands.

Preliminary experiments of this type have been carried out using crystals soaked with the **"P1" reaction products** phosphoryl-choline, -serine and -ethanolamine. In each case crystals survived treatment with relatively high concentrations of the ligand (up to 50mM) although they were more subject to radiation damage than is the native enzyme.

Difference Fourier maps for phosphorylcholine and phosphorylserine treated enzyme show the presence of a phosphate group located in the same position as P_i. This may imply hydrolysis of the phosphorylated ligands in the active site. The collection of an X-ray diffraction data set normally takes from 5 - 500 hours depending on the radiation source and the method used, a time scale several orders of magnitude greater than that for enzyme catalysed reactions.

A detailed examination of electron density maps for phosphorylcholine treated crystals also shows movement of amino acids in the active site of the protein, and high resolution data collection is now underway in the hope of detecting the whole ligand. NMR studies of the interaction of phosphorylcholine with PLC also seem to imply that the ligand is bound intact[38]. The situation with phosphorylethanolamine treated crystals is more complicated since it appears that, under certain circumstances, one of the Zn^{2+} ions in the active site is missing.

Treatment of PLC crystals with dicaproin (a **"P2" product** in the figure above) is subject to one of the major complicating factors in our present work, namely the insolubility of lipid and lipid fragments in the high salt conditions necessary to maintain stable crystals of the enzyme. Crystals are stable in the presence of droplets dicaproin for several days but finally disintegrate. We assume that this implies that dicaproin is absorbed into the crystals. We are planning further experiments.

Treatment of Crystals with Substrates ("S" above)

One molecule of PLC hydrolyses ca.50000 molecules of liposomal phospholipid[10] per minute. Activity is lower against monodisperse substrate but is still very rapid when compared with the time scale of a normal X-ray data collection. The almost total insolubility of the membrane lipids which are the normal substrates for PLC is also a major problem but we are hoping to overcome this using synthetic lipids with short chain fatty acid moieties.

Treatment of Crystals with Substrate Analogues ("S" above)

Phospholipase C activity is often assayed using p-nitrophenylphosphorylcholine (p-NPPC) as a substrate analogue[39]:

$$\text{p-NPPC} + H_2O \quad \xrightarrow[\text{Sorbitol}]{\text{PLC}} \quad \text{Phosphorylcholine} + \text{p-nitrophenol}$$

p-NPPC is a poor substrate (K_m = 0.2M) but p-nitrophenol has a deep yellow colour so that the reaction can be followed spectrophotometrically. The reaction requires the presence of sorbitol or glycerol.

p-NPPC can be soaked into crystals of PLC but the crystals yellow after exposure to X-rays and lose diffraction intensity much more rapidly than the native enzyme. This possibly due to hydrolysis of the substrate in the active site. Our data collection was therefore carried out at 7°C. Difference maps for the enzyme treated in this way show extensive changes in the active site of the protein and the presence of new electron density which may be due to the p-NPPC molecule or one of its reaction products.

Laue Methods[40, 41]

These are in principle a revival of the method which Friedrich & Kipping used in the first demonstration of X-ray diffraction and allow the very rapid collection of large amounts of X-ray data from protein crystals. The Laue technique utilises intense white X-radiation from a synchrotron source. The crystal seldom survives for more than a few minutes but often yields a virtually complete data set in the 2.5 - 5Å resolution range and on a time scale ranging from milliseconds to seconds. The data is recorded on a set of films which are digitalised using a micro-densitometer.

22

Mathematical methods for the reduction of the resulting raw data have been developed so that the technique is now being used quite extensively.

The Laue method can also be applied to crystals which are mounted in a "flow cell". Using this technique it is possible to collect data from a crystal of the "native" enzyme and then expose the crystal to a ligand and observe resulting structural changes as they occur by taking new exposures at short time intervals. Furthermore reactions may be initiated in crystals which have been pre-soaked with a ligand under inactive conditions by applying a sudden change to these conditions. Examples of this are temperature jump, pH jump, photodissociation of a "caged" substrate or the addition of co-factors or other effectors.

A preliminary time resolved study on PLC using the Laue/Flow cell technique and p-NPPC as substrate has shown an order- disorder-reorder phenomenon similar to that reported for phosphorylase B[40, 42]. This could lead to a detailed understanding of the reaction mechanism.

References

1.   Little,C. (1981) PLC. Methods in Enzymology 17, 725-730.

2.   Johansen,T., Holm,T. Guddal, P.H., Sletten, K. Haugli, F.B. & Little C. (1988), Gene 65,293-304.

3.   Guddal, P.H., Johansen, T., Schulstad, K., & Little, C. (1989), J. Bacteriol. 171, 5702 - 5706

4.   Little,C. (1977), Biochem. J. 175, 977-986.

5.   Little,C. & Johansen,S. (1979) Biochem. J. 179, 509-514.

6.   Sowadski,J.M., Hanschumaker,M.D., Krishna Murthy,H.M., Foster,B.A. & Wyckoff,H.W. (1985), J. Mol. Biol. 186, 417-433.

7.   Roberts,M.F, Otnaess, A.-B., Kensil,C.A. & Dennis,E.A. (1978), J. Biol. Chem. 253, 1252-1257.

8.   Aalmo, K.,Krane, J., Little, c. & Storm, C.B. (1984), Int. J. Biochem. 16, 931-934.

9.   Bicknell,R., Hanson,G.R. & Holmquist,B. (1986), Biochemistry. 25, 4219-4223.

10.   Little,C. (1977), Acta Chem. Scand. B31 267-272.

11.   El-Sayed,M.Y., DeBose,C.D., Coury,L.A. & Roberts,M.F. (1985), Biochim. Biophys. Acta 837, 325-335.

12.   Wolf,R.A. & Gross,R.W. (1985), J. Biol. Chem. 260, 7295-7303.

13.   Martin, T.W., Wysolmerski, & Lagunaff, D. (1987) Biochim. Biophys. Acta 917, 296 - 307

14.   Hostetler,K.Y., Hall, L.B. (1980) , Bichem. Biophys. Res. Commun. 96,388-393.

15.   Clark,M.A., Shorr,R.G.L. & Bomalski, J.S. (1986), Biochem Biophys. Res Commun. 140, 114-119.

16.   Besterman,J. M., Duronio,V & Cuatrecasas,P. (1986), Proc. Natl.Acad Sci. USA. 83, 6785 -6789.

17.   Levine,L., Xiaou, D-M. & Little,C. (1988), Prostaglandins 34, 633-642.

18. Hough,E, Hansen,L.K.,Birknes,B., Jynge,K., Hansen,S., Hordvik,A., Little,C., Dodson,E.J. & Derewenda, Z. (1989), Nature 338,357-360.

19. Brady,L.,Brzozowski, A.M.,Derewenda,Z.S., Dodson, E.,Dodson, D., Tolley, S., Turkenberg, J.P., Christiansen, L., Huge-Jensen, B., Norskov, L., Thim, L., & Menge, U. (1990), Nature, 343, 767-770).

20. Winkler, F.K., D´Arcy, A., & Hunziker, W. (1990), Nature, 343, 771-774).

21. Little,C. (1981), Acta Chem. Scand. B35 39-44.

22. Steitz,T.A., Beese,L., Freemont,P.S., Friedman,J.M. & Sanderson,M.R. (1987) , Cold Spring Symposia on Quantitative Biology LII, 465-471.

23. Blundell,T.L., Pitts,J.E., Tickle,I.J.,Wood, S.P. & Wu, C.-W. (1981), Proc. Natl. Acad. Sci. USA. 78,4175-4179.

25. Little, c. & Johansen, S. (1979), Biochem. J. 179, 509 - 514

26. Renetseder,R., Brunie,S., Dijkstra,B.W.,Drenth,J. & Sigler, P.P (1985), J. Biol. Chem. 260, 11627-11634.

27. Titball,R.W., Hunter,S.E.C., Martin,K.L., Morris,B.C., Shuttleworth,A.D., Rubidge,T., Anderson,D.W. & Kelly,D.C. (1989) , Infect. Immunity 57, 367-376.

28. Leslie,D., Fairweather,N., Pickard,D, Dougan,G. & Kehoe,M. (1989), Molec.Microbiol. 3, 383-392.

29. Yun Tso,J & Siebel,C. (1989), Infect. Immunity 57, 468-476.

30. Aakre, S.E. & Little, C. (1982) , Biochem. J. 203, 799-801.

31. Aalmo, K., Hansen, L., Hough, E., Jynge, K., Krane, J., Little, C. & Storm, C.B. (1984) , Biochem. Internat. 8, 27-33.

32. Hauser,H., Pascher,I., Pearson,R.H. & Sundell,S. (1981), Biochim. Biophys. Acta 650, 21-51.

33. Weaver,L.H., Kester,W.R., Matthews,B.W. (1977) , J. Mol. Biol. 114, 119-132.

34. Satow, Y., Cohen, G.H., Padlan, E.A. & Davies, D. (1986), J. Mol. Biol. 190, 593 - 604.

35. Burns, R.A. jr., Friedman, J.M., & Roberts, M.F. (1981), Biochemistry 24, 3100 - 3106

36. DeBose, C.D., Burns, R.A. jr., Donovan, J.M., & Roberts, M.F. (1986), Biochemistry 24, 1298 - 1306

37. Snyder,W.R. (1987), Biochim. Biophys. Acta 920, 155-160.

38. Anthonsen, T., University of Trondheim, (1990), personal communication.

39. Kurioka,S. & Matsuda, M. (1976), Anal. Biochem. 75, 281 - 289.

40. Hajdu, J. & Johnson, L.N. (1990), Biochem., 29, 1669-1678

41. Hajdu, J., Acharya, K.R., Stuart, D.I., Barford, D. & Johnson, L.N. (1988), Trends. Biochem. Sci., 13, 104-109).

42. Hajdu, J., McLaughlin, P.J., Helliwell, J.R., Sheldon, J. & Thompson, A.W. (1986), J. Appl. Crystallogr. 18, 528 - 532.

© *1991 Elsevier Science Publishers B.V. (Biomedical Division)*
Site-Directed Mutagenesis and Protein Engineering
M.R. El-Gewely, editor.

PDBASE :

POOR MAN'S STRUCTURAL PROTEIN DATABASE TOOL

by

Steffen B. Petersen
SINTEF MR-Center, N-7034 Trondheim, Norway.

INTRODUCTION

During the past 3 decades a steadily growing number of proteins have been studied using X-ray diffraction analysis. The main result of this effort has been the accumulation of detailed 3-dimensional data sets for these proteins. While the proteins are mainly water soluble and globular, we are still faced with serious difficulties if we want to study the 3-dimensional structure of membrane related proteins.

The available data encompass about 150 truly different high resolution structures, thus giving us atomic or near atomic resolution data for more than 30.000 residues in these proteins. The information presented in the data set is very primitive, basically consisting of a long list of the individual observed atoms together with their spatial location (x, y, z), a so-called "B"-factor describing the motion of the atom and an occupancy parameter describing the distribution of alternate conformations, if such is relevant.

Rational Protein Engineering is only possible if we use some sort of structural insight into the structure of the protein that we are aiming to modify the amino acid sequence of. The ideal situation is that we solve the 3D structure of the protein using X-ray diffraction analysis or NMR. Both of these methods are time consuming and one must anticipate a waiting period of at least a year before the 3D structure is available. Even when we have solved the native structure, we are still in a somewhat difficult situation, since we would like to be capable of predicting both the structural and functional consequences of the point mutations we are considering. If we can query all known structures for sequence and/or structural motifs we are in a much better position to recommend mutations that have a good chance of producing the changes we are trying to introduce into a given protein. In various laboratories around the world several projects are well underway aiming at developing professional tools for departmental computer systems.

We here describe a novel small database tool, that allows quite advanced queries to be made to this database of protein 3D structures . The program runs on a IBM PC or compatibles, and is very fast with execution times typically of the order of a few minutes on a Compaq 80386-20MHz.

METHOD

The PDBASE program is written in Microsoft "C" (version 6.0), it runs on IBM- PC's or compatibles – it does not require graphics or mathematical coprocessors although it will utilize a coprocessor in case one is present. The compiled program takes 63 Kb of storage space.

The data source used by the program is a processed version of the Brookhaven Protein Data Base (PDB). The "raw" data file is analyzed using the Kabsch & Sanders approach [1] leading to one new so-called DSSP file for every entry in the PDB. Each DSSP file contains an objective secondary structural assignment for every residue listed in PDB as well as the coordinates plus backbone torsion angles Φ, Θ, water accessibility and a few other parameters.

The secondary structure assignments together with the amino acid sequence is available to the PDBASE program in a single file containing sequence and secondary structure information on all PDB entries. The current size of this file is 182 Kb. The PDBASE program prompts the user for two search strings – one for the amino acid sequence and one (of identical length) for the secondary structure assignment. These strings may in part or totally be composed of wild cards (*). The program then searches both the sequence and secondary structure data base for any matches. In case of a dual match the program opens the corresponding DSSP's files and reads the appropriate local tertiary structural information and reports its results to the screen as well as to two files containing the results. Access to the DSSP files is optional, PDBASE will access them if they exist, and then continue searching the sequence and secondary structural file for further matches.

Several options exist for defining a match: the user can search for a certain level of similarity. In the case of similarity the program compares the amino acid types in the search string residue by residue with the more than 80.000 possible strings that can be extracted from the PDB. The user defines a threshold score above which the match criteria applies. The secondary structure string search only operates in perfect match mode, however the user can apply a simple logic using lower case and upper case letter e.g. capital letter H denotes the presence of a helical assignment at this position, whereas lower case letter h denotes the absence of helix at the location. Thus HHHHhhhh denotes an eight residue string containing one alpha-helix turn with 4 residues followed by four residues where there is not an alpha helix. This simple logical language turns out to be quite useful. One can search for specific motifs quite easily: a helix-turn-beta strand motif can be defined as HHHHhheeEE which defines a 4 residue turn separating an α-helix from a strand in a β-sheet conformation. One can also combine primary sequence and secondary structural search strings as is shown in example 1 (see below).

Finally, the user may apply a distance threshold criteria limiting the output to structures containing a match in terms of primary and secondary structure as well as displaying a certain distance between the first and the last residue in the search sequence. This simple feature allows the user to isolate loops with specific geometry, anti-parallel helices from other helix-helix configurations.

EXAMPLE 1: TRYPTOPHANES TERMINATING α-HELICES

In this example we have requested PDBASE to report back all occurrences of tryphtophane at the very end of α-helical stretches. Our search sequences are :

```
SEQUENCE  STRING :  ***W****
STRUCTURE STRING :  HHHHhhhh
```

LABEL	VER:	SEQUENCE	SEC.STRUCT
1CHG	3:	KKYWGTKI	HHHHTT__
1HDS	1:	KAAWGKVG	HHHHTTTS
1HHO	2:	TALWGKVN	HHHHTT_
2CGA	1:	KKYWGTKI	HHHHGGG_
2CHA	3:	KKYWGTKI	HHHHGGG_
2CPP	1:	QEAWATLQ	HHHHGGGG
2DHB	1:	KAAWSKVG	HHHHTTSG
2HCO	2:	TALWGKVN	HHHHTT_
2HHB	2:	TALWGKVN	HHHHTT_
2HHB	4:	TALWGKVN	HHHHTT_
2MHB	2:	LALWDKVN	HHHHTT_
3BCL	3:	RBFWFIGP	HHHHSTTT
3HHB	2:	TALWGKVN	HHHHTT_
4APE	1:	SAYWAQVS	HHHHTTST
4CHA	3:	KKYWGTKI	HHHHGGG_
4CHA	6:	KKYWGTKI	HHHHGGG_
4CTS	1:	SKEWAKRA	HHHHTTS_
4CTS	2:	SKEWAKRA	HHHHTTS_
4FD1	1:	AEDWDGVK	HHHHTT_S
4HHB	2:	TALWGKVN	HHHHTT_
4HHB	4:	TALWGKVN	HHHHTT_
5CHA	3:	KKYWGTKI	HHHHGGG_
5CHA	6:	KKYWGTKI	HHHHGGG_
6CHA	3:	KKYWGTKI	HHHHGGG_
6CHA	6:	KKYWGTKI	HHHHGGG_
CONSENSUS	:	KKYWGKVI	HHHHTTGG

Figure 1: *A listing produced by PDBASE of α-helices with a terminating tryptophane. The column 'LABEL' contains the Brookhaven Data Bank entry code for the protein in which the match was found. The column 'VER' is indicating the chain number in DSSP containing the dual match. The 'SEQUENCE' and 'SEC.STRUCT' columns contain the sequences and secondary structures that matched the search criteria. In the secondary structure string H denotes helix, G 3-10 helix, B bridge (a loop feature), E extended strand (β), S a bend in the backbone exceeding 15° and T turn.*

As seen in figure 1 based on this query profile the PDBASE program has isolated 27 8-membered amino acid sequences which are α-helical in the first 4 residues whereas the last four are not α-helical. In addition the strings reported back all contain tryptophane at the C terminal of the helix. PDBASE also informs that only 10 out of the 27 matches were having different sequences. Finally it calculates which residues were most often reported at any given position in the sequence and reports this as the 'consensus' sequence.

EXAMPLE 2: PROLINES IN α-HELICES

In figure 2 is shown how the prolines are distributed in the α-helices found in PDB. The results strongly suggests that no prolines are to be expected in the α-helices of globular soluble proteins in the second turn or further downstream. Thus if the structural principles deducted from a study of globular proteins can be extended to membrane bound as well as structural proteins, we must expect a strong "kink" e.g. in trans membrane helices, if a proline is present in the membrane spanning stretch.

The PDBASE is also capable of evaluating this postulate, by quering whether we can find helix-gap-helix stretches where the proline is placed in or close to the gap. Performing this particular search, one finds that proline is always defining the onset of the second helix and surprisingly enough is not figuring as the terminating residue of the first helix (not shown here).

28

Figure 2 : Distribution of prolines in α-helices.
Along the horizontal axis is shown the α-helical position monitored for the presence of a proline. N +1 indicates the first residue in the α-helix in the N-terminal, N +2 the second and so forth. Likewise C –1 is the last residue in the α-helix which actually terminates the helix.

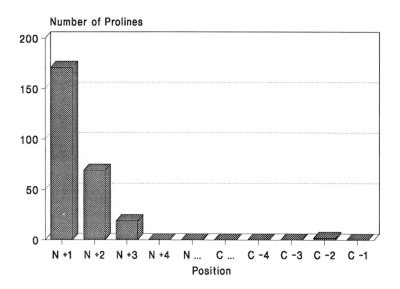

EXAMPLE 3:   PROTEASE SEARCH

It is often relevant to identify sequence and structure for homologous proteins. A very nice example of the abilities of PDBASE is given in figure 3. The search sequence has been selected from Subtilisin Carlsberg in the surrounding of the active HIS-64. No search criteria was given for the secondary structure string. Several Subtilisin-like structures as well as their associated secondary structural assignments are identified in the output from PDBASE. Obviously we are here dealing with a highly conserved region of the sequence. This is not the case for a similar stretch of amino acids isolated from the trypsin class of proteins, where the divergence is much more pronounced.

If we want to search for all subtilisin-like sequences which are similar but not exactly identical to the sequence selected from 2SBT (subtilisin Carlsberg) we set up our search criteria as follows :

```
SEQUENCE  STRING : SHGTHVAGT
STRUCTURE STRING : *********
```

The output of the program contains a listing of all occurrences that matched the combined search criteria. Please note that the program searches for *sequence similarity*.

```
        LABEL : SEQUENCE              SEC.STRUCT

        1CSE :  GHGTHVAGT             SHHHHHHHH
        1SBC :  GHGTHVAGT             SHHHHHHHH
        1SBT :  SHGTHVAGT             SHHHHHHHH
        1SEC :  GHGTHVAGT             SHHHHHHHH
        1SNI :  SHGTHVAGT             SHHHHHHHH
        2PRK :  GHGTHCAGT             SHHHHHHHH
        2SBT :  SHGTHVAGT             _HHHHHHHH

CONSENSUS :     GHGTHVAGT             SHHHHHHHH
```

Figure 3 : *Protease sequence search. PDBASE finds several subtilisin like protease sequences when querying with a sequence segment selected close to the active HIS. For further details see legend of Figure 1.*

EXAMPLE 3:   MOTIF SEARCH

The PDBASE program allows for a search for the occurrence of rather long motifs. As an example one can search for an α-turn-β motif with the aim of isolating those hits that contain the two structural features in a parallel motif. Since PDBASE as part of its output gives the C-α – C-α distance between the first and the last residue one can separate different folds by simply sorting according to this distance.

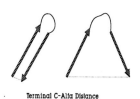

Terminal C-Alfa Distance

Figure 4 : *PDBASE allows the user to define a distance threshold   between the terminal C-α atoms, which then subsequently is used for screening the DSSP files. The user thus is capable of differentiating between densely packed motifs and more open motifs.*

CONCLUSION

A simple but versatile tool has been discussed and illustrated. It is believed that tools similar in nature will be extremely relevant for protein engineering research as well as for protein folding studies. Since the PDBASE program is running on IBM-PC and compatibles and only is producing ASCII output it should be installable on most PC's without problems.

We are offering this program as well as the sequence and secondary structure database for free to academic laboratories. In order to obtain a copy, please send a diskette compatible with your system and an envelope with your proper postal address. Our only demand is that any publication including utilization of the program should make reference to the program, as well as to the author.

(1) Kabsch, W and Sander, C. (1983) FEBS Letters 155, 179.

© 1991 Elsevier Science Publishers B.V. (Biomedical Division)
Site-Directed Mutagenesis and Protein Engineering
M.R. El-Gewely, editor.

Electrostatics of Proteins

P. Martel[1] and S.B. Petersen[1,2]

[1]SINTEF MR-Center, N-7034 Trondheim, Norway.
[2]Dept. of Chemistry, Århus Universitet, DK-8000 Århus, Denmark.

Introduction

Electrostatics effects in proteins have shown to be of importance in a number of phenomena like titration behaviour [1], molecular recognition and binding [2], structure stabilization [3], and enzyme mechanisms [4].These effects are due to the presence of ionizable groups, partially charged polar groups and polarized bonds. Moreover, the protein atoms are polarizable, meaning that they can give rise to multipoles by the action of an external electric field [5, 6].

The first electrostatic model of a protein was introduced by Linderstrøm-Lang in 1924 [7], in an attempt to explain the titration behaviour of albumin. The protein was represented as a sphere with charges smeared over its surface. This first attempt was too simplistic to fully account for the experimental data. In 1957 Tanford and Kirkwood [8] introduced a markedly better model, where the protein was described as a low dielectric "cavity" immersed in a high dieletric solvent. An analytic solution for the Poisson-Boltzmann (PB) equation was presented, for the case of a spherical cavity.The PB equation plays a leading role in protein electrostatics, since it gives as a solution the potential in a system where the dielectric constant and ionic strength may vary through space. In the Tanford-Kirkwood model the charges are not randomly distributed over the protein surface, but placed at specific locations. The lack of structural data at that time led Tanford and Kirkwood to study various hypothetical charge distributions. Later, when the first X-ray structures of proteins were available, Tanford and Roxby [9] included structural information in their electrostatic calculations, but without success since the structural data applied in their model was not enough.

Shire et al. [10] used static accessibilities [11] for the side groups together with the Tanford-Kirkwood theory, leading to markedly better results. This model was extensively used [12, 13, 14], and proved to be quite satisfactory in the modeling of titration curves. The main flaws of this approach are the use of a spherical boundary and the absence of the self-energy term in the energy calculations [15] (the self-energy of a charge in a dieletric cavity is the potential energy of the charge in its reaction field, due to the uneven polarizabilities of the two dieletric materials [16]).

The first numerical algorithm for solving the Poisson equation in proteins was presented by Warwicker and Watson [17], although this problem had previously been discussed by Orttung [18]. The Warwicker and Watson algorithm was used to study enzyme-substrate interaction [19], structure stabilization by α-helix dipoles [20], potential changes at the heme iron of cytochrome c1 [21], and pKa shifts in subtilisin [22] .

Numerical solution of PB equation was first achieved by Klapper et al [23] and the algorithm was later improved by Honig et al [24], giving quite good results in a number of problems [25, 24, 26, 27]. The commercial program DELPHI implements this algorithm, featuring solution of both the linear and non-linear forms of the PB equation.

The main problem with finite-difference methods (such as numerical solution of PBE) is the amount of time required for the calculations. Distributing the charge over the grid points, the discrete approximation of the protein-solvent boundaries, and the large errors observed close to the protein-charges are also serious problems.

All the procedures described so far fall into the class of *continuum models*, which model the protein and solvent regions as continuous media described by dieletrics constants D_p and D_{solv}. A different class of methods, the so called *microscopic models*, attempt to model both the effects of the permanent dipoles of the surrounding water molecules and the induced dipoles created by the action of the permanent charges upon the electron clouds. In order to evaluate the induced dipoles, atomic polarizabilities are needed, and estimates can be obtained from quantum mechanics. The microscopic approach has been extensively used by Warshel and Russel [6]. Warshel's algorithm is available in a software program called POLARIS, part of the MOLARIS modeling package.

Finally the surface potential method, from Zauhar and Morgan [28, 29], calculates the distribution of the polarization charge density over the molecular surface, resulting from the set of charges in the protein. The electric potential can then be computed at every point. The main problem with this method is the representation of the complex surface of a protein as a collection of small curvilinear elements [29].

Despite the existence of more detailed models as described above, the use of a simple constant dieletric model is still attractive for some applications, such as visualization of electrostatic potential rather than force, and if qualitative observations suffice.

Materials and Methods

In this work we used the simplest of the continuum methods, namely the *constant dieletric* method [5]. In this approach the protein-solvent region is modeled as an uniform medium of dieletric constant D. The charges are therefore immersed in this medium and the potential at a point P is given by:

$$\phi_P = \frac{\sum_i^N q_i}{4\pi\epsilon_0 D d}$$

where the summation extends over the whole set of N charges in the protein. These include both the ionized groups and the permanent dipoles along covalent bonds. Partial atomic charges are taken from the CHARMM molecular mechanics program [30]. Arg and Lys residues are assigned charges values of +1; Glu and Asp residues are assigned charges of -1. This all or none charge assignment cannot account for the titration behaviour and is a crude approximation of the charge distribution at pH 7, in particular for residues that titrate close to the pH value in question, such as His.

The electrostatic potential is computed for every intersection point of a cubic grid containing the protein molecule and a fraction of the surrounding solvent. Surfaces of constant potential are displayed for different energy levels, measured in units of kT.

The grid step used is 2.0 Å, and the grid dimensions 50X50X50. With this setting it takes about 20 minutes to perform a calculation on a Personal Iris 4D-20 workstation (running at 2 MFLOPS).

The potential was computed and visualized by means of graphics software tool called EMAP written by the authors.

We focused our attention on the family of trypsin-like proteases [31]. Crystal coordinates are available in the PDB database [32] for some of the members of this family, namely trypsin, α- and β-chymotrypsin, elastase and kallikrein. Trypsin-like proteases belong to the broader family of *serine proteases*. Serine proteases seem to exhibit an universal catalytic mechanism, dependent

Trypsin-like proteases					
Name	Code[a]	Charge[b]	Specificity	Optm. pH	Source
Trypsin	1SGT	-1.0	Lys-X, Arg-X	8.0	*S.griseus*
Elastase	3EST	+3.0	Hydrophob	8.5	Porc.Pancreas
Chymotrypsin	4CHA	+2.0	Hydrophob	7.5	Bov.Pancreas
Kallikrein	2PKA	-18.0	-	7-8	Porc.Pancreas

[a]Brookhaven Protein databank entry code
[b]Arg and Lys with charge +1, Glu and Asp with charge -1, all other residues neutral

Table 1: Some characteristics of the trypsin-like proteases used in this work.

Homology data				
-	1SGT	3EST	4CHA	2PKA
1SGT	-	10.17	15.00	9.94
3EST	10.17	-	22.04	16.51
4CHA	15.00	22.04	-	19.14
2PKA	9.94	16.51	19.14	-

Table 2: Homology scores obtained with the PIR alignment program. Scores are the distances in standard deviation units between the similarity score of the 2 sequences and the average score of the randomized sequences. Parameters used: bias 6, break penalty 6, # of random runs 60.

on the presence of the residues Asp, His and Ser at the active site, the so called *catalytic triad*. The details of the enzymatic mechanism are yet to be understood, but electrostatic effects seem to play a leading role [4].

Despite marked structural similarities and significant sequence similarities as well (table 2), trypsin-like proteases show different preferences for the cleavage points (table 1). This variability could be explained by the different residues lining the active pocket [31].

In the present work we compare the electrostatic potentials surfaces of these 4 proteins - trypsin, chymotrypsin, elastase and kallikrein.

Results and Discussion

The EMAP program was used to calculate the potential fields for the 4 selected trypsin-like proteases. For each protein, surfaces of constant potential were computed, for several different kT values. Positive surfaces are displayed in cyan, and negative surfaces in blue. The zero potential surface (when shown) is colored white. Figures 1a-f show the results for bacterial trypsin, pancreatic elastase, chymotrypsin and kallikrein.

On figure 1a the zero potential field surface for a stretch of polypeptide chain is shown, as calculated by EMAP. Notice how the zero potential surface can be complex, even in a small piece of a protein molecule.

In figure 1b the potential energy levels -1, 0 and 1 kT were displayed for bacterial trypsin. The catalytic triad (His 57, Asp 102 and Ser 195) is displayed using a Van der Waals surface

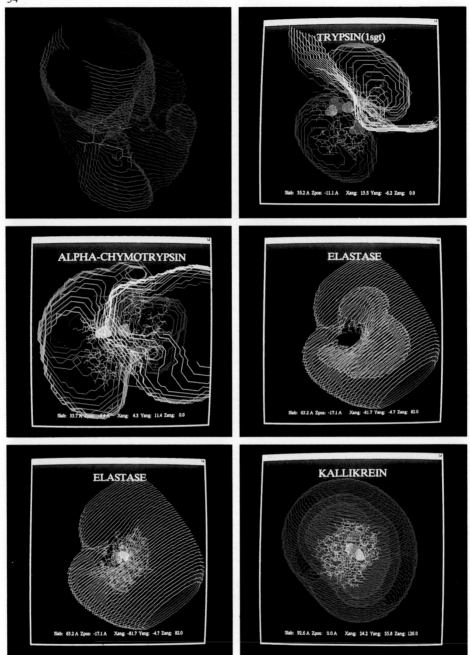

Figure 1: Electrostatic potential surfaces (positive surfaces are shown in blue, negative surfaces in red and the zero potencial surface in white): a) polypeptide, 0.0 kT   b) trypsin, -1.0,0.0,1.0 kT   c) chymotrypsin, -1.0,0.0,1.0 kT   d) elastase, 0.5,1.0 kT   e) elastase, 1.0 kT   f) kallikrein, -3.0,-2.0,-1.0 kT

representation (colors yellow, red, magenta). The negatively charged Asp 189, lying in the bottom of the active site pocket, is also highlighted in red, using VDW surfaces. The dipolar nature of the protein is easily seen, with the active site lying close to the zero level potential surface. The active pocket is placed on the negative site, which correlates well with the fact that this enzyme shows a preference to accommodate a positively charged side group (Arg or Lys) in the active pocket.

The calculated surfaces for chymotrypsin (figure 1c) show the same dipole structure seen in trypsin. The active site pocket is again lying close to the zero level surface.

The electrostatic field of elastase is shown in figure 1d, using potential surfaces at +0.5 and +1.0 kT. Note how the active site closely matches the "hole" in the positive surfaces, figure 1e.

The calculated potential surfaces for kallikrein are shown in figure 1f. The strong negative potential (surfaces shown are -3, -2, -1 kT) is due to the high number of negatively charged side groups (Asp and Glu) present in this protein. At pH 7.0, the net charge of kallikrein is -18.0, not including possible counterions. In this case it is not possible to correlate the observed potential surfaces with structural features, due to the rather "atypical" nature of the potential field. This molecule could be stabilized by counterions or by a positively charged substrate.

The results presented here, although somewhat preliminary in their nature, show a tendency for a dipole structure in the trypsin-like proteases. The active site is placed close to the center of this dipole, near the zero energy potential surface.

With the exception of kallikrein, all the proteins studied are close to the isoelectric point at pH 7.0. This suggests that a calculation should be done for kallikrein at its isoelectric point, by using a titration procedure to find the charge distribution at the isoelectric pH. This facility is currently being implemented in EMAP.

Acknowledgments
Paulo Martel thanks Junta Nacional de Investigação Científica e Tecnológica for his grant.

References

[1] Matthew, J.B. (1985) Ann.Rev.Biophys.Biophys.Chem. 14:387-417

[2] Warwicker, J. (1989) J.Mol.Biol. 206:381-395

[3] Rogers, N.K. (1989) in "Prediction of protein structure and the principles of protein conformation" Fasman, G.D.(ed.) Plenum Press, NY 1989 pp.359-389

[4] Warshel, A., G.Naray-Szabo, F.Sussman and J.-K.Hwang (1989) Biochemistry 28(9):3629-3637

[5] Rogers, N.K. (1986) Prog.Biophys.Mol.Biol. 48:37-66

[6] Warshel, A. and S.T.Russel (1984) Q.Rev.Biophys. 17:283-422

[7] Linderstrøm-Lang, K. (1924) C.R.Trav.Lab.Carlsberg 15:70

[8] Tanford, C. and J.G.Kirkwood (1957) J.Am.Chem.Soc. 79:5333-5339

[9] Tanford, C. and R.Roxby (1972) Biochemistry 11:2192

[10] Shire, S.J., G.I.H.Hanania and F.R.N.Gurd (1974) Biochemistry 18:1919-1939

[11] Lee, B. and F.M.Richards (1971) J.Mol.Biol. 55:379-400

[12] Matthew, J.B., G.I.H.Hanania and F.R.N.Gurd (1979) Biochemistry 18:1919-1939

[13] Matthew, J.B. and F.M.Richards (1982) Biochemistry 21:4989-4999

[14] Matthew, J.B et al (1985) Crit.Rev.Biochem. 18:90-197

[15] Warshel, A., S.T.Russel and A.K.Churg (1984) Proc.Natl.Acad.Sci. USA 81:4785-4789

[16] Friedman, H.L. (1975) Mol.Phys. 29:1533-1543

[17] Warwicker, J. and H.C.Watson (1982) J.Mol.Biol. 157:671-679

[18] Orttung, W.H. (1977) Ann.N.Y.Acad.Sci. 303:22-37

[19] Warwicker, J. (1986) J.Mol.Biol. 121:199-210

[20] Rogers, N.K. and M.J.E.Sternberg (1984) J.Mol.Biol. 174:527-542

[21] Rogers, N.K., Moore, G.R. and M.J.E.Sternberg (1985) J.Mol.Biol. 182:613-616

[22] Sternberg, M.J.E, F.R.F.Hayes, A.J.Russel, P.G.Thomas and A.R.Fersht (1987) Nature 380(5):86-88

[23] Klapper, I., R.Hagstrom, R.Fine, K.Sharp and B.Honig (1986) Proteins 1:47-59

[24] Gilson, M., K.Sharp and B.Honig (1987) J.Comp.Chem 9:327-335

[25] Gilson, M. and B.Honig (1987) Nature 330:84-86

[26] Gilson, M. and B.Honig (1988) Proteins 3:32-52

[27] Sharp, K., R.Fine and B.Honig (1987) Science 236:1460-1463

[28] Zauhar, R.J. and R.S.Morgan (1985) J.Mol.Biol. 186:815-820

[29] Zauhar, R.J. and R.S.Morgan (1988) J.Comp.Chem. 9(2) 171-187

[30] Brooks, B.R. et al (1983) J.Comp.Chem. 4:187-217

[31] Schulz, G.E. and R.H.Schirmer "Principles of protein structure" Springer NY pp. 183-186

[32] Brookhaven Protein Data Bank entry codes: 1SGT, 3EST, 4CHA, 2PKA.

APPLICATIONS OF MASS SPECTROMETRY IN BIOTECHNOLOGY

JOHN S. SVENDSEN

University of Tromsø, Institute of Mathemathical and Physical Sciences,
Department of Chemistry, N-9000 Tromsø, Norway.

INTRODUCTION

Historically, mass spectrometry (MS) has been a unique technique in combining extreme accuracy with very high sensitivity, albeit the scope of the technique has been limited to volatile compounds of low molecular weight. The molecules of biotechnology; proteins, peptides, and nucleic acids, are all non volatile compounds with very high molecular weights. The advent of a new generation of mass spectrometers with new ionization methods for large and involatile compounds incorporating recent technology for mass measurements in high mass area has changed the situation and initiated a large number of applications of MS in the biosciences, also in biotechnology (1-3).

In principle, the mass spectrum provides two kinds of information; the molecular weight of the compound which can be deduced from the molecular ion, and secondly, structural information derived from the fragment ions. Peptides and nucleic acids are both classes of compounds quite suited for structural analysis by mass spectrometry, being polymers of a relatively limited set of monomers. It should be emphasized that not all techniques in mass spectrometry provides both types of information; fast atom bombardment (FAB) ionization, which is the premier ionization method for biomolecules, yields mass spectra with prominent molecular ions, while fragment ions are not formed, or formed with low intensity. In some of the new mass spectrometers, this inherent limitation can be circumvented by multi stage (tandem) mass spectrometry (MS/MS) (4,5). In this technique the molecular ions formed by FAB ionization are focussed in the first mass spectrometer and subsequently decomposed to fragments in a collision cell, before a final mass analysis in the second mass spectrometer. The fragments formed by the collision are often of great structural value, e.g. in peptides the fragmentation takes place at or around the peptide bonds. The beauty of this method is that it works for all kinds of posttranslational modifications (6) as well as cyclic peptides (7) as opposed to Edman degradation. The penality is is greatly diminished sensitivity, often a thousand times as much materials are required relative to FAB molecular weight determination. Another problem with this procedure is that several amino acids have identical or similar mass, and cannot thus be distinguished.

One should be aware, however, that in many practical applications a molecular weight determination is sufficient to resolve a practical problem. We will in the following, present two examples which illustrates some of the possibilities in biomolecule research provided by mass spectrometry.

RESULTS

The first example is a verification of a questionable amino acid composition. The compound in question is a cyclic peptide, Kalatapeptide B, with uteroactive properties, isolated from an African plant (*Oldenlandia affinis DC*) (8). The amino acid composition was determined as shown in Table 1.

Table 1. Proposed Amino Acid Composition for Kalatapeptide B (from ref. 8)

Amino Acid	Amount	No.	Amino Acid	Amount	No.
Asp	2.01	2	Thr	4.86	5
Ser	1.10	1	Glu	1.09	1
Pro	3.97	4	Gly	4.91	5
Cys	5.96	6	Val	2.73	3
Leu	1.00	1	Trp	1-	1
Arg	0.97	1	TOTAL		30

This amino acid composition gives a calculated molecular weight of 2991.77, under the assumtion that the peptide is cyclic and that all cysteins forms disulfide bonds. FAB-MS of this peptide shows a protonated molecular ion at 2894, which corresponds to a molecular mass nearly 99 Dalton below the expected value. Closer inspection of the amino acid composition reveals that the measured number of valine residues deviates quite significantly from the integer number assigned for Kalatapeptide B. The mass of one Val residue is 99 Dalton, and an interpretation of these results gives the conclusion that the number of valines in the peptide is only 2, not 3 as suggested by amino acid analysis, thus the total number of residues are only 29.

The example above shows that the determination of molecular weight can verify or falsify a proposed amino acid composition. There is in this experiment not produced any information with regard to the molecular structure (amino acid sequence). As previously stated this information is only available by the formation of fragments. We are, however, not bound to make fragments inside the mass spectrometer, fragments can as easily be produced outside by chemical or enzymatic cleavage of the peptide. This technique can be applied on proteins/peptides useing endopeptidases or specific chemical cleavage reagents, and perform molecular weight mapping analysis, allowing readily detection of altered primary structure (as a result of genetic engeneering) (9). Another alternative is to use exopeptidases and perform C- or N-terminal sequencing. This latter method requires a continuous sampling of the reaction mixture, a technique available through the dynamic FAB method, where the FAB probe tip (where ionization takes place) is connected to the reaction vial by a capillary. When a peptide sample is treated with an exopeptidase, the probe tip will continuously be infused by the shortened peptides as they are formed by terminal degradation of the starting peptide. Each of these shortened peptides will produce a molecular ion and by following the formation of the different molecular ions as the degradation proceeds, the mass of the residues which is removed in each step can be detected. An example of dynamic FAB sequencing is illustrated by the verification of the C-terminal sequence of a synthetic human IgG(Fab) fragment by treatment with carboxypeptidase Y and P. The results are presented in table 2.

Table 2. Determination of the C-terminal sequence of the synthetic human IgG(Fab) fragment.

Time (min)	M1 m/z=1856	M2 m/z=1728	M3 m/z=1671	M4 m/z=1558
0	100	100	18	12
4	6	56	66	13
7	0	20	100	45
20	0	5	13	100

The mass difference between M1 and M2 corresponds to a loss of an Lys, between M2 and M3 to Gly, and between M3 and M4 to Leu/Ile, thus establishing the C-terminal sequence as Lys-Gly-Leu/Ile (C-terminal to N-terminal). The sequence could be followed 4 more residues.

CONCLUSION

This presentation has shown only a tiny fraction of the possibilities offered by mass spectrometry in the analysis of biomolecules. A more complete survey is given in the references found in the introduction. It should, however, be clear that mass spectrometry is a very flexible approach to obtain the type of information needed for different problems. The determination of the molecular weight can be sufficient to correct an amino acid composition, or establish that a certain modification in the primary structure has beeen achieved by mapping analysis. Structural information can be gained by the formation of fragments either inside the mass spectrometer (MS/MS) or by combining enzymatic degradation with dynamic FAB.

MATERIALS AND METHODS

The experiments were performed on a VG Analytical TRIBRID multi stage mass spectrometer equipped with a Cs LSIMS (liquid secondary ion mass spectrometry - ion FAB) ion source. Static FAB were performed by dissolving the peptide in 25 % TFA in water. This solution were applied on glycerol matrix on the FAB probe tip. The dynamic FAB experiment was performed as outlined in the literature (10). The mass spectra were recorded in positive ion mode using multi channel analysis, summing spectra for one min. The mass spectrometer were calibrated using CsI.

REFERENCES
1   Busch KL, Cooks RG (1982) Science 118:247-254
2   Rinehart Jr. KL (1982) Science 118:254-260
3   McEwen CN, Larsen BS (eds.) (1990) Mass Spectrometry of Biological Materials, Marcel Dekker, New York
4   Biemann K (1990) in McEwen CN, Larsen BS (eds.) (1990) Mass Spectrometry of Biological Materials, Marcel Dekker, New York pp 3-24
5   Hunt DF, Shabanowitz J, Yates JR, Griffin PR, Zhu NZ (1990) in McEwen CN, Larsen BS (eds.) (1990) Mass Spectrometry of Biological Materials, Marcel Dekker, New York pp 169-196
6   Carr SA, Roberts GD, Hemling ME (1990) in McEwen CN, Larsen BS (eds.) (1990) Mass Spectrometry of Biological Materials, Marcel Dekker, New York pp 87-136
7   Eckart K, Schwarz H, Tomer KB, Gross ML (1985) J. Am. Chem. Soc. 107:6765-6769
8   Sletten K, Gran L (1973) Meddelelser fra Norsk Farmaceutisk Selskap 35:69-82
9   Morris HR, Dell A, Panico M, Thomas-Oates J, Rogers M, McDowell R, Chatterjee A (1990) in McEwen CN, Larsen BS (eds.) (1990) Mass Spectrometry of Biological Materials, Marcel Dekker, New York pp 137-168
10   Caprioli RM (1987) Mass Spectrom. Rev. 6:237-287

© 1991 Elsevier Science Publishers B.V. (Biomedical Division)
Site-Directed Mutagenesis and Protein Engineering
M.R. El-Gewely, editor.

Neural Networks applied to the study of Protein Sequences and Protein Structures

H. Fredholm[1,2]
and
H. Bohr[4], J. Bohr[3], S. Brunak[4],
R.M.J. Cotterill[4], B. Lautrup[2] and S.B. Petersen[2]

[1] Niels Bohr Institute, Blegdamsvej 17, DK-2100 København Ø, Denmark.
[2] SINTEF MR-Center, N-7034 Trondheim, Norway.
[3] Risø National Laboratory, DK-4000 Roskilde, Denmark.
[4] The Technical University of Denmark, B. 307, DK-2800 Lyngby, Denmark.

Introduction

Within the last two decades X-ray crystallographers have solved the 3-dimensional (3D) structures of a steadily increasing number of proteins in the crystalline state, and recently 2D-NMR spectroscopy has emerged as an alternative method for small proteins in solution. These methods have so far solved the 3D structure of close to three hundred proteins, but only about half of these can be regarded as truely different. The number of protein sequences known today is well over 10,000, and this number seems to be growing at least one order of magnitude faster than the number of known 3D protein structures. It is therefore of great importance that we develop tools that can predict structural properties of proteins from their sequence.

Neural Networks

Recent developments in artificial neural networks [1] provide a new promising computational technique for analyzing protein sequences and protein structures. The term artificial neural networks, or just neural networks, here refers to massively parallel computer programs that employ brain-inspired computational strategies, see Figure 1.

A neural network performs a task in terms of mapping an input pattern of reals or binary numbers to an output pattern of reals or binary numbers. By presenting to the neural network examples of a given input-to-output mapping, the network can, because of its remarkable ability to discriminate between patterns, learn to map the input patterns to the output patterns for the examples shown. Furthermore, when given an input pattern it has never seen before, the neural network uses its ability to generalize and makes a prediction of what the output pattern should be. The quality of this prediction will depend on the similarity between the input pattern and the patterns the network was trained on.

Thus, in contrast to conventional programming techniques where a program is specified by explicitly giving an algorithm that describes step by step how the program should compute, a neural network is "programmed" in terms of examples. If given sufficiently many examples of correct input-output patterns, the neural network will extract the implicit algorithm defined by the examples—even if no known algorithm exits. It is exactly this property that makes neural networks an interesting tool for studying protein sequences and protein structures.

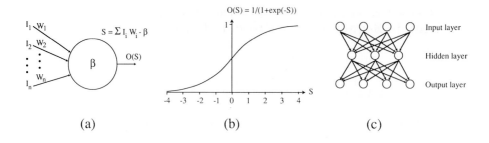

$$S = \Sigma I_i W_i - \beta$$

$$O(S) = 1/(1+\exp(-S))$$

(a)   (b)   (c)

Figure 1: *a)* The basic building block in a neural network. The neuron receives input from other neurons over a set of input connections $I_1, \ldots, I_n$. Each of these inputs has a value between 0 and 1. The neuron weighs these inputs with a specific weight for each input connection, adds the resulting values together and subtracts an internal threshold β. If the resulting value s ($s = \sum_{i=1}^{n} I_i w_i - \beta$) is below zero the neuron emits a value close to zero, if s is above zero the neuron emits a value close to one and if s is very close to zero the neuron emits a value around 0.5, see *b)*. The actual function is $o = 1/(1 + \exp(-s))$. *c)* A neural network is formed by organizing the neurons in three layers called input layer, hidden layer and output layer. Information flows only "forward" (*i.e.* downward in the figure) from input layer to output layer. Input is fed into the network at the input layer by assigning a value between 0 and 1 to each neuron in this layer. Each neuron in the input layer then sends its assigned value to all neurons in the hidden layer, which compute their own output in parallel according to the rule defined in *a+b)*. When the outputs for all the neurons in the hidden layer have been computed, each neuron in the hidden layer sends its output to the neurons in the output layer. The neurons in the output layer compute their output in the same way as the hidden neurons did. The result of the networks computation is the output present at the output layer. It should be noted that it is the weights and the threshold of each neuron at the hidden layer and the output layer (the neurons at the input layer have no weights or threshold) that determine the result of the computation for a given input. The neural network is programmed by setting these weights and thresholds. The weights are not set manually. There exist a learning algorithm [1] that can find a set of weights and thresholds that will make the network map a given set of input-to-output examples correctly.

Primary Structure Homology

A neural network can be used to measure sequence homology between two proteins. This is accomplished by learning the neural network to predict the identity of a residue from the context the residue appears in, see Figure 2a. The homology between two protein sequences A and B is then computed as how good a network trained on sequence A is to predict sequence B, see table 1.

It turns out that this measure complies with the PIR [2] measure if we restrict PIR from using insertions and deletions in the alignment. However, if PIR is allowed to make insertions and deletions in the alignment, then PIR scores significantly better than the neural network on sequences with low homology. In contrast to PIR the neural network can also measure homology to a family of proteins since it can be trained on a set of proteins.

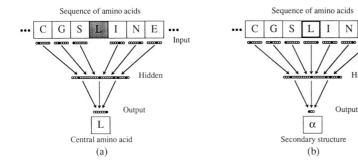

Figure 2: *a)* Primary homology network. The network has fifty input groups with twenty neurons in each group representing a stretch of seventeen contiguous residues in the protein (only six groups and seven neurons/group are illustrated). Each group code for one residue by activating one of the twenty neurons. The output layer has twenty neurons. The neuron that takes the highest value wins. *b)* Secondary structure network. This network is similar to the homology network, except that the central residue is now shown to the network and the output layer has three instead of twenty neurons, *i.e*, one for each secondary category.

Primary Structure Prediction by Neural Network										
	3RP2	2EST	2KAI	1CHG	4PTP	2TRM	2PRK	1SAV	1SEC	2SBT
3RP2	100.0	19.6	12.1	14.9	15.7	18.8	6.8	9.2	8.4	9.4
2EST	14.3	100.0	13.8	16.7	20.2	17.5	9.7	9.3	9.9	11.6
2KAI	14.8	10.8	100.0	11.8	22.4	22.9	2.9	6.3	4.1	5.5
1CHG	11.6	15.8	13.8	100.0	23.3	19.3	8.2	7.1	9.9	8.3
4PTP	17.0	18.3	21.1	23.2	100.0	72.6	9.3	11.5	6.6	6.9
2TRM	20.5	15.8	24.1	21.9	70.9	100.0	7.9	9.3	5.8	5.8
2PRK	5.4	7.9	3.4	7.9	9.0	6.7	100.0	13.8	13.1	18.2
1SAV	9.4	5.8	3.4	8.8	8.1	9.0	18.3	100.0	48.2	43.3
1SEC	8.9	7.1	4.3	11.0	10.3	6.7	15.1	46.5	100.0	61.5
2SBT	7.1	4.2	7.3	7.5	7.2	4.5	17.6	43.5	59.5	100.0

Table 1: Homology as determined by neural network with an input window of fifty residues and forty neurons in hidden layer. The protein names refer to the Brookhaven Protein Data Bank entry codes. The first column list the protein sequence the network was trained on. The other columns list the homology scores in terms of percent correctly predicted residues. The table is almost symetric, *e.g*, training the network on one 4PTP and predicting 2PRK gives the same result as training on 2PRK and predicting 4PTP. Note also that the trypsins (3PR2–2TMR) and the subtilisins (2PRK–2SBT) have a high score ($\gtrsim 12\%$) within their group and a low score ($\lesssim 10\%$) outside.

Secondary Structure Prediction by Neural Networks						
Group	Window	Hidden	Output	Proteins	Score (%)	Type of Prediction
Qian *et al.*	17	40	3	108	64	helix, sheet, coil
Bohr *et al.*	51	40	2	56	72	helix, not helix
Holley *et al.*	17	2	2	48	63	helix, sheet, coil

Table 2: Architecture and performance of different neural networks. The window column gives the number of residues in the input window, the hidden and output columns give the number of neurons in the hidden layer and the output of the network and the protein column gives the number of proteins used. The score column gives the stated performance for those proteins that were not used in the training set. The method applied by Bohr *et al.* is expected to give rise to higher scores due to the lower number of output catagories. Thus the scores are not directly comparable (table adapted from [7]).

Secondary Structure Prediction

The neural network used for secondary structure prediction is very similar to the network used for measuring primary homology, see Figure 2b. The network is typically taught the task of classifying each residue in a protein according to whether its secondary structure is part of a helix, sheet or coil, where coil represents everything that is not helix or sheet.

Several groups [3, 4, 5] have used this method for predicting protein secondary structure and all results reported conform in quality with the results achieved by traditional statistical methods, although the neural network method seems to be up to 10% better. All protein structure prediction schemes do, however, suffer from a poor statistical basis for claimed prediction quality, due to the small number of proteins the prediction method is based upon and the even smaller number of proteins that are used for measuring the prediction quality. Table 2 summarizes the best results reported for neural networks and list some of the major differences in their approach.

Tertiary Structure Prediction

Recently, a novel approach to prediction of the 3D structure of protein backbones based on a neural network similar to the network shown in Figure 2b was put forth [6]. In addition to training the network to produce a secondary structural assignment, the network was also trained to reproduce binary distance constraints within a sequence distance of 30 residues, *i.e.*, if the C_α atom of another residue was less than a given threshold distance away a 0 was assigned, whereas if further away a 1 was assigned. A large scale network with 61 amino acids in the input window, 300 hidden units, and 33 output units (3 for secondary structure and 30 for binary distance constraints) was trained. For proteins not used during training, but homologous to those used in the training the network produced reasonably accurate binary distance constraints. Using as initial guess the backbone conformation for a homologous protein, the authors reported a case with only 3 Å rms deviation from the coordinates determined by X-ray diffraction.

Understanding Insertions and Deletions

We have found that there exist a linear correlation between the neural network measure for primary homology and neural networks ability to predict protein secondary and tertiary structure (unpublished results).

Our study of the homology network showed us that while the homology network is very robust towards mutations it is sensitive to insertions and deletions. Due to the linear correlation between the prediction ability of the networks, we believe this to be true for secondary and tertiary structure prediction as well.

In order to understand insertions and deletions [as understood by neural networks] we have looked at a prediction scheme that combines three networks. One network bases its prediction on the context to the left of the central residue only, another bases its prediction on the right context only and a third network looks at both contexts. The reason for using networks that look at the left or right context only was that we hoped that one of these networks would be able to produce the right prediction, when the full-context network failed because of an insertion or deletion in the protein sequence.

To decide between the three predictions the networks produced, we used a winner-network-take-all strategy, that is, the prediction of network with the highest value on one of its output neurons was used.

An optimal decision strategy would never select a network that makes a wrong prediction in favor of a network that makes the right prediction. The winner-network-take-all strategy do, however, sometimes make such selections.

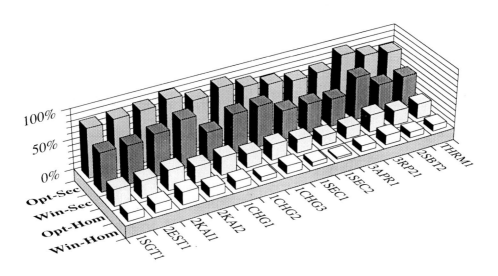

Figure 3: Results for combined networks. The training set consisted of 4PTP, 2PRK, 1SCG, 2ALP and 2TRM.

Figure 3 shows the results obtained with the winner-network-take-all strategy and compares it to an optimal decision strategy. Both homology and secondary structure predictions are shown. Three things should be noted:

- The prediction ability for primary homology and secondary structure is highly correlated.

- The winner-network-take-all strategy is far from optimal. A high value on one of the output neurons is not always the best measure for how confident the network is in its prediction—

the values on the other output neurons should at the same time be low. Another problem is that often one network has 0.7 as higest value and another has 0.8 as highest value, the former being correct while the latter being wrong.

- An optimal decision strategy will give very high prediction scores. This indicates that the networks that looks at the left or right context only are capable of making right predictions when the full-context network fails. Provided that a decision strategy for selecting the right network could be deviced the prediction ability of neural networks could be increased with up to 20% on secondary structure prediction giving a prediction score between 80–90%.

The performance of winner-network-take-all strategy was comparable to the performance of the full-context network in the experiment reported here. New experiments [8] with networks that predicted whether a residue is part of an α-helix or not showed, however, that the winner-network-take-all strategy was 5% better than the full context network predicting on its own.

Conclusion

Neural networks provide a new tool for studying protein homology, secondary protein structure and tertiary protein structure. The homology measure defined by neural networks complies with traditional measures, except from being more sensitive to insertions and deletions. On secondary structure prediction neural networks perform up to 10% better than traditional methods. The method for tertiary structure prediction has been applied with success in a single case study. Further studies are need in order to evaluate this method. Combination of several different networks seems to be able to improve the prediction on secondary structure with up to 20%, provided a good strategy for deciding which network to believe can be deviced. The currently best strategy can give an improvement of approximately 5%.

References

[1] Rummelhart, D.E. *et al.* (1987) Parallel Distributed Processing, Vol 1: Foundations, MIT Press.

[2] The Protein Identification Resource (PIR) "ialign" program with the unitary protein matrix (bias = 6 and penalty = 1000) was used.

[3] Qian, N. and Sejnowski, T.J. (1988) J. Mol. Biol. 202, 865-884.

[4] Bohr, H., Bohr, J., Brunak, S., Cotterill, R.M.J., Lautrup, B., Nørskov, L., Olsen, O.H. and Petersen, S.B. (1988) FEBS Letters 241, 223-228.

[5] Holley, L.H. and Karplus, M. (1989) Proc. Natl. Acad. Sci. 86, 152-156.

[6] Bohr, H., Bohr, J., Brunak, S., Cotterill, R.M.J., Fredholm, H., Lautrup, B. and Petersen, S.B. (1990) FEBS Letters 261, 43–46.

[7] Bohr, H., Bohr, J., Brunak, S., Cotterill, R.M.J., Fredholm, H., Lautrup, B. and Petersen, S.B. Neural Networks and Biological Sequences, Trends in Biotechnology, November 1990.

[8] Brunak, S. Private communications, unpublished results.

Acknowledgements
HF thanks the Danish Research Academy, Novo-Nordisk and UNI-C for grants.

PROTEIN ENGINEERING

© 1991 Elsevier Science Publishers B.V. (Biomedical Division)
Site-Directed Mutagenesis and Protein Engineering
M.R. El-Gewely, editor.

PROTEIN ENGINEERING

DALE L. OXENDER* AND THOMAS J. GRADDIS**

Department of Biological Chemistry, University of Michigan Medical
School, Ann Arbor, Michigan, USA 48109-0606. **SmithKline Beecham,
King of Prussia, Pennsylvania, USA 19404-2799

Protein engineering is a result of the fusion between recombinant DNA technologies and recent advances in physical biochemical
techniques.[1,2,3] The primary goal of protein engineering is the
rational design and construction of novel proteins with enhanced
or unique properties. The immediate challenge for the field of
protein engineering is to understand the rules or code relating the
primary sequence of a protein to its three-dimensional structure
and then to determine how that structure is related to function.
The enormous economic power of protein engineering to enhance
current use of proteins and introduce novel uses is propelling the
development of this field at an astonishing rate. Protein engineering will profoundly extend our understanding of the molecular
detail of the chemical world as well as produce products for the
present and future fields of agriculture, chemical industry, and
medicine.

The successful application of protein engineering requires an
interdisciplinary approach dependent on a flow of information between the five areas of expertise outlined in Figure 1. An examination of the efforts currently being carried out suggests that
studies of well-established model systems that lend themselves to
an interdisciplinary collaboration will be most effective for
developing the generic tools for protein engineering.

Enabling Technologies: The cycle of protein engineering

Structural analysis. The heart of protein engineering lies
in obtaining a broad range of structural information so that structural features can be related to specific function. Obtaining the
three-dimensional structure of a protein is usually the rate-
limiting step in the cycle outlined in Figure 1. In addition, it
is essential to acquire structural information of an engineered
protein when testing design hypotheses so that the new structural

---

*Vice President, Biotechnology, Warner-Lambert Company, 2800
Plymouth Road, Ann Arbor, Michigan 48105

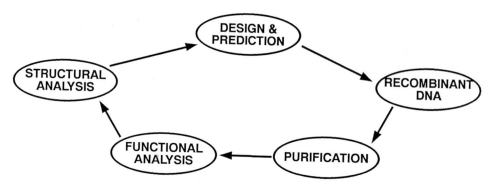

Fig. 1. The cycle of Protein Engineering. Schematic representation of the five major paradigms comprising the field of protein
engineering and the relationship between them.

elements can be related to altered function. Crystallographic
techniques for determining protein structure are rapidly improving.
The intrinsic weakness of a crystallographic approach is that suitable protein crystals must be obtained. Two-dimensional Nuclear
Magnetic Resonance (NMR) spectroscopy is a powerful technique for
examining the structure of a polypeptide in solution. This method
bypasses the difficult crystallization step in X-ray diffraction.
The major limitation to NMR is that it currently has been useful
only for small proteins of mass less than 30,000 daltons.

Design and prediction. From knowledge of the relationship
between structure and function, predictions and design principles
(next circle, Fig. 1) can be used to develop plans for modifying
the structure of the protein. Mutational studies presently play
a central role in identifying sites within protein structures
critical to function. It is premature to say we are designing
proteins based on theoretical predictions, rather we are modifying
existing structures in nature to test hypotheses.

Knowledge based predictions represent one of the current
approaches to design. Modeling the three-dimensional structure
of a protein can be based on sequence homologies of other proteins
of known structure, usually proteins within the same functional
family. As a result of these models, successful design of renin
inhibitors has led to valuable pharmaceutical products.

Effective application of protein engineering necessitates a
coupling of experimental data with theoretical studies for

predicting the folded structure of proteins and for the development
of models that anticipate the conformational effects of a given
amino acid substitution. Energy minimization and molecular dynamic
procedures are constantly being improved and have exhibited considerable success in predicting protein secondary structure, yet
their limitations painfully remind us that the prediction of tertiary structure remains a major bottleneck for progress in protein
design.

Realizing the full potential of protein engineering will require rational de novo design and, therefore, it remains as an
important long-range goal. Several successful examples of de novo
protein design and construction involve only small polypeptides
whose structures are entirely composed of either beta-sheet or
alpha-helices. A recent significant advance is the design of a
chymotrypsin-like polypeptide molecule that exhibits significant
catalytic activity.

Recombinant DNA. The isolation of a gene coding for a protein, by cloning or by synthetic methods, is a prerequisite to
protein engineering. The cloned gene will serve as the template
for engineering the protein product. For production of polypeptides in cells, the efficiency of gene expression is affected by
both the structure of a particular gene and the physiology of the
host cell. Production of many proteins in bacteria can be advantageous since they can usually be scaled-up to produce large quantities in a relatively short time at a modest cost. Production of
mammalian proteins in mammalian cell culture overcomes the problem
of post-translational modification associated with bacterial production. An alternative approach to the production of mammalian
proteins is the expression of these proteins in the milk of mammals where they are processed and modified as fully functional
proteins for human medicinal use.

Purification and functional analysis. The challenge of protein purification is to isolate one protein from this complex
mixture with reasonable efficiency, yield and purity, while keep-
ing in mind that for each step of purification the yield decreases
and the cost of producing the protein increases. It is often
possible to use specific properties engineered into proteins to
greatly assist their expression and purification.

Specific and sensitive assays of protein function are essential for evaluation of the results of site-directed mutation steps, often allowing the investigator to bypass structural analysis of the mutant protein. Protein engineering has been used to facilitate cascade or coupled protein assays and immunological diagnostics.

Progress report in protein engineering

In the agricultural field, increasing the biological value of plant proteins is a target for protein engineering. Also, specific growth hormones have been engineered that result in accelerated growth of animals and plants. The medical field, including biomedical research, acute and chronic patient care, disease prevention, and medical instrumentation, already uses many products of protein engineering. Since secondary structure of intermediate-sized hormones appear to dominate receptor-hormone binding interactions, the design of intermediate-sized polypeptide hormones with enhanced biological activity is fast becoming a reality. The near-term products of protein engineering will encourage the interface between the worlds of biology and microelectronics. Proteins are being used to construct biosensor electronic devices that utilize the unique catalytic and molecular recognition properties of proteins.

Considerable success has been achieved in altering the thermal, chemical, and biological clearance stability of several important proteins. Proteins can often be profoundly stabilized by combining a number of selected mutations in one molecular. Changing the substrate specificity of an enzyme may involve changing only a few residues, or it can require the wholesale movement of an entire domain. It should be kept in mind that virtually any application presently using proteins can likely be profoundly improved through protein engineering. As a result of protein engineering, entirely new enzymatic processes are being developed for cheaper, faster, and less hazardous chemical manufacture, waste treatment, and biomass conversion.

REFERENCES

1. Oxender, DL, Fox CF (1987) Protein Engineering. Alan R. Liss, Inc., New York.

2. Oxender, DL (1985 and 1987) Protein Structure, Folding, and Design, I and II. Alan R. Liss, Inc., New York.

3. Netzer WJ (1990) Biotechnology 8:618-622

© 1991 Elsevier Science Publishers B.V. (Biomedical Division)
Site-Directed Mutagenesis and Protein Engineering
M.R. El-Gewely, editor.

HIGH-LEVEL EXPRESSION OF BOVINE PANCREATIC RNase A

KATRIN TRAUTWEIN, STEVEN A. BENNER

Laboratory for Organic Chemistry, E.T.H. Zurich 8092 Switzerland

INTRODUCTION

In 1984, Nambiar et al. reported the first example of a chemically synthesized gene adapted to "cassette" or "modular" mutagenesis (1), a gene for ribonuclease (RNase) from bovine pancreas. Unique restriction sites were place throughout the gene to facilitate its cutting and pasting, and therefore the preparation of mutant genes. Since this work, many others have followed this simple idea to create genes suited for site-directed mutagenesis. We have recently published a computer program that can aid in the design of synthetic genes suitable for protein engineering (2). Further, we have used our synthetic gene for RNase and its descendents (including a gene for angiogenin)(3) to make the first reconstructed proteins from ancient organisms (4), to make hybrids between RNase A and its homologue angiogenin (5), and to introduce point mutations that help us better understand the catalytic mechanism, substrate specificity, and quaternary structure of RNase A. Especially exciting in this work has been the emergence of evidence that extracellular ribonucleases, and their substrates, extracellular RNA molecules, play a role as intercellular messages in higher organisms (5,6).

However, the easiest part of site-directed mutagenesis is the preparation of the mutant genes themselves. Most of the work (and virtually all of the science) comes later, when these mutant genes are expressed and isolated, and the effects of the point mutation are interpreted in terms of structure. A consensus has not yet emerged to dictate how much structural information must be collected on a mutant protein before reliable interpretations can be drawn. On one hand, point mutations that greatly alter the structure of a protein are presumed to be rare. On the other hand, it is by no means clear what a "great" alteration is in this context. Catalysis is clearly influenced by displacement of the catalytic groups by as little as a half of an Angstrom. Thus, virtually all intepretations of the behavioral impact of a point mutation are complicated should the overall conformation of a protein be altered by a point mutation even by an amount that is considered to be "small" (<1.5 Å) from a crystallographer's point of view.

It is, of course, possible to do an x-ray crystallographic analysis of every mutant that is analyzed. However, the emergence of high-resolution NMR methods (7) offers an alternative approach, provided that certain technical difficulties can be overcome. The most serious limitation of high-resolution NMR experiments for determining the structure in proteins (7) is the overlap of signals arising from different protons in a protein. Resonances that cannot be resolved cannot be assigned. This limitation means that NMR methods have to date been applied almost exclusively to proteins whose molecular weights are less than ca. 10,000.

The introduction of a third frequency domain (8) into a multidimensional NMR experiment offers a solution to the problem that arises from overlap of resonances, and promises to extend the molecular weight range of proteins amenable to NMR analysis (9,10). Three dimensional NMR can also allow direct analysis of structural perturbations created by point mutation, especially if isotopic labeling (e.g. ^{15}N) is introduced into the protein. It is not unreasonable to expect that we will shortly be able to make a relatively complete set of statements regarding the conformational alteration resulting from a point mutation can be derived from a single heteronuclear three-dimensional NMR spectrum.

RNase has, of course, been studied extensively by classical NMR techniques (11-13). However, its molecular weight and stable structure make it ideal as a system to illustrate the use and test the sensitivity of heteronuclear ^{15}N-proton 3D-NMR technology. Before heteronuclear ^{15}N-proton 3D-NMR can be routinely applied to mutants of RNase, however, milligram quantities of ^{15}N-labeled protein must be routinely available. This, in turn puts significant demands on these systems for heterologously expressing RNase in *E. coli* cells. Several of the

methods developed in these laboratories (14) are promising, but
the optimal strain of *E. coli* for expressing RNase is in most
cases an auxotroph for several amino acids, unable to synthesize
these amino acids from ammonia present in the growth medium;
these amino acids must be supplied as such for the organism to
grow. This means that a protein expressed in these hosts grown
on (relatively inexpensive) ^{15}N-labeled ammonia will not contain
^{15}N in the amino acids for which it is an auxotroph. Supplying
^{15}N-labeled amino acids directly (e.g., ^{15}N-lableled histidine)
is, of course, prohibitively expensive.

Therefore, we have developed a new protocol for expressing
large amounts of bovine pancreatic RNase A labeled with ^{15}N
using M9-type media (15). The key step in this procedure is the
growth of wild type *E. coli* on minimal medium labeled with ^{15}N
(as ammonium chloride) to yield, in ^{15}N-enriched cells, a
complete mixture of amino acids fully labeled with ^{15}N. This is
then used as the protein supplement for the growth of the
auxotrophic expression host. We suspect that this method will be
useful for other workers wishing to express large amounts of
heterologous proteins labeled with ^{15}N and suitable for 3D-NMR
work. Further, in this work, we found that the codons used in
the 5'-end of the message had a significant impact on the level
of expression of the protein, information that will again be
useful generally to those wishing to express large amounts of
protein in *E. coli*.

MATERIALS AND METHODS
Preparation of Mutants

Oligonucleotides were synthesized on an Applied Biosystems
model 380A synthesizer using phosphoramidite chemistry and
purified either by plyacrylamide gel electrophoresis (PAGE) on
16% gels as described by Applied Biosystems in the users manual,
or by silica gel chromatography using as eluant
butanol:ammonia:water (55:35:10).

The double stranded oligonucleotides introduced into the
gene by modular mutagenesis had the following sequences:

```
5'-CATG AAA GAA ACT GCA GCA GCA AAA TTT GAA AGA CAA CA
         TTT CTT TGA CGT CGT CGT TTT AAA CTT TCT GTT GTAT-5'


5'-CATG AAA GAA ACT GC
         TTT CTT TGA CGC CGG-5'
```

Complementary oligonucleotides were annealed,
phosphorylated with kinase, and ligated according to procedures
described in Maniatis et al.(15). The vector pJLA502 was
purchased from Medac GmbH. The gene for RNase was obtained from
a synthetic gene (1,16), subcloned into pUC19 (16). *E.coli*
strains W3110, RB791 and lon⁻ (kindly provided by Dr. G.
McGeehan) were transformed by calcium shock (15), and plasmids
obtained from transformations were isolated and sequenced.
Protein was quantitated by UV (ε_{280} =0.73 for 1 mg/ml RNase) and
confirmed using the Bradford (17) protein assay supplied by Bio-
Rad.

Expression and Purification of RNase

^{15}N medium was prepared by innoculating M9 medium (15) made
with ^{15}NH$_4$Cl with E.coli strain W3110, letting this grow to
maximum density, harvesting the cells by centrifugation and
saving the spent clarified medium. The cell paste was taken up
in an equal volume of 12 N HCl and heated to 110°C under vacuum
for 10 hours. The HCl was then removed by evaporating the
solution to dryness twice under reduced pressure, and the dark
residue taken up in a small volume of the spent medium. This
solution was filtered, diluted with the remaining clarified
medium, adjusted to pH 7.4 with NaOH, and autoclaved. Prior to
innoculation, glucose, antibiotics and salts were added as for
M9 medium.

^{15}N medium (25 ml) containing ampicillin only was
innoculated with a lon⁻/pUN2-RNase from a glycerol culture, the
culture allowed to grow until the doubling time reached approx.
75 minutes, then 5 ml of this culture used to innoculate 450 ml
^{15}N medium in an erlenmeyer flask (2 l). As soon as this culture
reached an optical density (at 550nm) of 1.0-1.2 at 30°C, it was

transferred to a shaking water bath preheated to 42°C and expression was allowed to continue for 2.5 hours. The cells were recovered by centrifugation (15 min at 5000 x g), then the paste was suspended in 4 volumes (v/w) pH 7.8 STE buffer (50 mM NaCl, 50 mM Tris, 1 mM EDTA, 1 mM PMSF). The resuspended cells were lysed by passage through a French press at 8000 psi, the insoluble fraction recovered by centrifugation (15 min at 12000 x g) and solubilized in 9 volumes (v/w) denaturing buffer (8M urea, 100 mM NaCl, 50 mM Tris, 1 mM EDTA, 50 mM 2-mercaptoethanol, pH 7.8). After completely suspending the pellet, this solution was diluted with 9 volumes redox buffer (100 mM NaCl, 50 mM Tris, 1 mM EDTA, 1 mM PMSF, 10 mM glutathione reduced, 1 mM glutathione oxidized, pH 7.8). The solution was stirred at 4°C overnight, centrifuged to remove sedimented proteins (15 min at 10000 x g), then applied to a Sephadex G25 column equilibrated with 20 mM NH_4OAc pH 7.0. The protein-containing fractions were loaded onto a CM-Agarose column (Bio-Rad), the column washed with 10 volumes 20 mM NH_4OAc and eluted using 10 column volumes of a linear salt gradient from 20 to 2M ammonium acetate, pH 7.0. The fractions containing UpA activity were pooled, desalted by repeated concentration and dilution into 50 mM NaOAc pH 5.0 using an Amicon concentrator with a YM10 membrane, and stored at 4°C.

RESULTS

An expression vector was constructed from commercially available pJLA502, which contains the strong leftward and rightward bacteriophage lambda promoters P_L and P_R in tandem (18), the synthetic gene for RNase (1), and synthetic linkers from the AUG containing Nco I recognition site into the gene. A significant increase in the intensity of the RNase band on SDS-PAGE could be obtained by using a linker lacking the Not I restriction endonuclease recognition site near the 5'-end of the message. Using protease-deficient E.coli strain lon⁻, which lacks heat-inducible protease La and has a mutation in the locus that regulates the heat shock response of E.coli, hptR (19,20), more ribonuclease could be isolated from the cell paste than when using strains RB791 or JM101. Ribonuclease with universal

[15]N-labelling could be isolated in yields of 2-3 mg/L medium. Using LB medium (15), 5-15 mg/L ferment could be obtained.

The purification procedure involved removal of all soluble proteins following cell lysis, solubilization of insoluble proteins including RNase with urea, renaturation, and ion-exchange chromatography on CM-agarose resin (21). Contaminating proteins (if any) could be removed by a second CM-agarose column, yielding essentially homogeneous product with 100% specific activity as judged by SDS-PAGE.

DISCUSSION

An expression system for pancreatic ribonuclease A allowing great flexibility in choice of a host E.coli strain is described above. The amount of RNase expressed could be increased by a factor of 10 solely by altering the N-terminal sequence of the gene to remove a Not I site.

Either secondary structure or codon usage may account for the change. The mRNA for the gene containing the Not I site is predicted by the program Squiggles (Wisconsin Data Package) to give a stable base-paired stem between codons 4 and 5 (GCG GCC) and the Shine-Dalgarno sequence (GGAG) plus translation initiation region. Mutating the codons to yield the modified gene (codons 4 and 5 become GCA and GCA) disrupts this base-pairing, and the Shine-Dalgarno sequence is predicted to be in a loop. Similar disruption of a stable hairpin between the Shine-Dalgarno sequence and codons 3-5 of the interferon mRNA was found to account for a 4-fold increase in translational efficiency (22).

Alternatively, it is worth noting that the codon initially used for Ala 5, GCC, is rarely used in highly expressed genes in E.coli (23-25). The GCA codon, in the mutated gene, is more commonly found in highly expresed proteins. Altering Leu 2 of the mouse thymidylate synthase gene from CUG to CUU increased the level of expression by a factor of 17 (26), whereas a 4-fold increase in the expression levels of tetanus toxin fragment C was obtained by Makoff et. al. by removing all rare codons (27).

These experiments do not permit a firm conclusion regarding why the mutation here has such a strong effect on the level of expression of RNase. However, it points to the need to structurally vary the messenger RNA to achieve high levels of expression, and may explain, at least in part, difficulties other laboratories have had expressing high levels of genes for various RNases (28).

The growth medium, essentially M9 medium enriched with hydrolyzed bacteria, was found to give substantially more labelled protein than M9 medium, a result that was consistent with the reduced doubling times and higher cell densities obtained on the enriched M9 compared to M9 medium alone.

The combination of an efficient expression system, a protease-deficient host strain, a rapid purification procedure, relatively inexpensive labelled medium, and newly developed multi-dimensional NMR technology should be applicable to other proteins as well, and should serve to broaden the scope of proteins amenable to NMR studies beyond those of rather low molecular weights.

ACKNOWLEDGEMENTS

We are indebted to the Swiss National Science Foundation and Sandoz AG for support, and to Dr. Joseph Stackhouse for valuable discussions.

REFERENCES
1. Nambiar KP, Stackhouse J, Stauffer D, Kennedy WP, Eldredge JK Benner, SA (1984) Science 222:1299-1301

2. Presnell SR, Benner SA (1988) Nucl Acids Res 16:1693-1702

3. Strydom DJ, Fett JW, Lobb RR, Alderman EM, Bethune JL, Riordan JF, Vallee BL (1985) Biochem 24:5486-5494

4. Stackhouse J, Presnell SR, McGeehan GM, Nambiar KP, Benner SA (1990) FEBS Lett 262:104-106

5. Benner SA, Allemann RK (1989) Trends Biochem Sci 14:396-397

6. Benner SA (1988) FEBS Lett 233:225-228

7. Wüthrich K (1986) NMR of Proteins Nucleic Acids, Wiley, New York

8. Griesinger C, Sorensen OW, Ernst RR (1987) J Magn Reson 73:574-579

9. Marion D, Kay LE, Sparks SW, Torchia DA, Bax A (1989) J Am Chem Soc 111:1515-1517

10. Vuister GW, Bodens R, Padilla A, Kleywegt GJ, Kaptein R (1990) Biochem 29:1829-1839

11. Alonso J, Paolillo L, D'Auria G, Nogues V, Cuchillo CM (1989) Int J Peptide Protein Res 34:66-69

12. Rico M, Bruix M, Santoro J, Gonzalez C, Neira JL, Nieto JL, Herranz J (1989) Eur J Biochem 183:623-638

13. Knoblauch H, Rüterjans H, Bloemhoff W, Kerling KET (1988) Eur J Biochem 172:485-497

14. McGeehan GM, Benner SA (1989) FEBS Lett 247:55-56; Nambiar KP, Stackhouse J, Presnell SR, Benner SA (1987) Eur J Biochem 163:67-71

15. Maniatis T, Fritsch E, Sambrook J (1982) Cold Spring Harbor Laboratory, Cold Spring Harbor, NY

16. Yanisch-Perron C, Vieira J. Messing J (1985) Gene 33:103-119

17. Bradford M (1976) Analytical Biochem 72:248-254

18. Shauder B, Blöcker H, Frank R, McCarthy J (1987) Gene 52:279-283

19. Goff SA Goldberg AL (1985) Cell 41:587-595

20. Baker TA, Grossman AD Gross CA (1984) Proc Natl Acad Sci USA 81:6779-6783

21. Taborsky G (1959) J Biol Chem 234:2652-2656

22. Spanjaard RA, van Djik M, Turion AJ, van Juin J (1989) Gene 80:345-351

23. McPherson DT (1988) Nucleic Acids Res 16:4111-4120

24. Ikemura T (1985) Mol Biol Evol 2:13-34

25. Grosjean H, Fiers W (1982) Gene 18:199-209

26. Zhang H, Cisneros RJ, Dunlap RB, Johnson LF (1989) Gene 84:487-491

27. Makoff AJ, Oxer MD, Romanos MA, Fairweather NF, Ballantine S (1989) Nucleic Acids Res 17:10191-10202

28. Miranda R (1990) Dissertation, Massachusetts Institute of Technology

© 1991 Elsevier Science Publishers B.V. (Biomedical Division)
Site-Directed Mutagenesis and Protein Engineering
M.R. El-Gewely, editor.

MOLECULAR ANALYSIS OF LEUCINE-BINDING PROTEIN SPECIFICITY

DALE L. OXENDER AND MARK D. ADAMS
Department of Biological Chemistry, University of Michigan, 1301 E. Catherine St.,
Ann Arbor, MI 48109

INTRODUCTION

Escherichia coli contains two kinetically distinguishable osmotic-shock sensitive transport systems for leucine. These two systems, LIV-I and LS, exhibit different substrate specificities based on the participation of two different periplasmic binding proteins. The two binding proteins, LIV-BP and LS-BP, interact with a common set of membrane-bound components which are shared by the LIV-I and LS systems to complete the translocation of leucine across the bacterial inner membrane (2).

Binding studies and assays of amino acid transport have shown that LIV-BP recognizes leucine, isoleucine, valine, and to a lesser extent threonine, serine and alanine (3). LS-BP is specific for leucine of the naturally occurring amino acids, but also binds the leucine analog 5′5′5′-trifluoroleucine (TFL) with high affinity (3). Other branched chain amino acid analogs have also been reported as inhibitors of leucine uptake, including: azaleucine (4), 2-aminobicyclo[2,2,1]heptane-2-carboxylic acid (BCH) (5), and others (6, 20).

The genes encoding LIV-BP and LS-BP have been cloned and sequenced (7). The proteins are about 80% identical at the amino acids level. X-ray diffraction studies of both proteins in the unliganded conformation were reported by Sack, *et al.* (8, 9) at a resolution of 2.8Å. For LIV-BP, the authors also examined the structure after soaking the unliganded crystals in a leucine solution. The free leucine associated with several residues in the N domain of the protein, leading to a prediction of a leucine binding site (8).

The carboxy and amino groups of the bound leucine molecule participate in a large array of side chain and peptide backbone hydrogen bond interactions that provide effective binding without counter-charged ions. All of these protein residues are identical in LS-BP. The hydrophobic side chain of the leucine molecule bound to the LIV-BP crystals interacts with six residues: Tyr_{18}, Leu_{77}, Cys_{78}, Ala_{100}, Ala_{101}, and Phe_{276} (Figure 1). Three of these, Leu_{77}, Cys_{78}, and Ala_{101}, are conserved in LS-BP, whereas the other three, Tyr_{18}, Ala_{100}, and Phe_{276}, are different in LS-BP. We have examined the contribution of each of these latter three residues to ligand specificity.

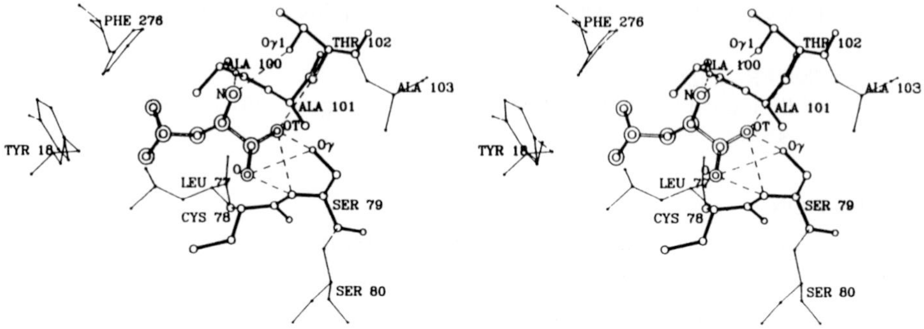

Figure 1. Stereo diagram of the leucine binding site of LIV-BP. From Sack, *et al.* (8). The leucine binding site was determined by soaking unliganded crystals of LIV-BP in a leucine solution prior to collecting the X-ray diffraction pattern. A difference electron density map revealed extra density associated with the bound leucine. All residues are conserved in LS-BP except Tyr$_{18}$, Ala$_{100}$, and Phe$_{276}$, which are Trp, Gly, and Tyr, respectively.

EXPERIMENTAL PROCEDURES

Bacterial strains and plasmids. See Table 1.

Site-directed mutagenesis. Performed by the method of Kunkel (10) as modified by Su and El-Gewely (11) for use with Sequenase™. Breifly, the gene to be mutated was cloned into bacteriophage M13 and the recombinant phage was passaged through an *E. coli* strain which allows incorporation of deoxyuridine into the DNA. An oligonucleotide complementary to the *livK* or *livJ* gene and containing the desired mutation was synthesized and hybridized to single-stranded, deoxyuridine-containing bacteriophage DNA. The modified phage T7 DNA polymerase Sequenase™ was used to synthesize the second DNA strand with the mutagenic oligonucleotide as the primer. Following ligation, the double-stranded DNA was transformed into wild-type *E. coli*, which select against the deoxyuridine containing strand and in favor of the strand containing the mutation. Mutations were identified by DNA sequencing.

Binding protein purification. LB broth (12) supplemented with 100 µg/ml ampicillin was inoculated 1:100 with a fresh overnight culture of *E. coli* strain

Table 1. Bacterial Strains and Plasmids

Strain Name	Source/Genotype
DH5αF'	λ⁻ *thi-1 gyr*A *rel*A1 *hsd*R17(r$_K$⁻,m$_K$⁺) *sup*E44 *rec*A1 *end*A1 F'φ80d*lacZ*ΔM15(*lacZYA-arg*F)U169
AE510	(7)
BL21(DE3)	(18)
Plasmid Name	**Source/Construction Notes**
pKSty	pOX7 (21) with StyI restriction site at codon 124 of *livK*
pJSty	pOX15 (22) lacking the *htpR* region (HindIII-PstI fragment) and containing a StyI restriction site at codon 124 of *livJ*
pKWY18, etc.	mutant derivatives of pKSty
pJWY18, etc.	mutant derivatives of pJSty
pJK	hybrid of pKSty and pJSty carrying codons -23 to +124 of *livJ* and +125-346 of *livK*
pKJ	hybrid of pKSty and pJSty carrying codons -23 to +124 of *livK* and +125-344 of *livJ*
pT7-5	(17)
pT7KSty, pT7JSty, etc.	*livJ*, *livK*, etc. cloned into pT7-5
pTKHMGF	(23)
pKStyHMGF	HindIII-BamHI fragment from pKSty in pTKHMGF
pKGA100HMGF	HindIII-BamHI fragment from pKGA100 in pTKHMGF

BL21(DE3) harboring a pT7-derived plasmid which carried a binding protein gene. The cells were grown to OD$_{600}$=1.0. Transcription by T7 RNA polymerase was induced by addition of isopropylthiogalactoside (IPTG) to a final concentration of 0.4 mM followed by three more hours of growth. Cells were then pelleted by centrifugation and submitted to the cold osmotic shock procedure (13). The shock fluid was dialyzed overnight against several changes of 10 mM potassium phosphate buffer, pH 6.9. Total protein was usually around 2.5 mg/ml. Leucine-binding protein represented seventy to ninety per cent of the total shock fluid protein.

Binding protein assay. Binding activity was measured by equilibrium dialysis either directly using radiolabeled ligands (K$_D$ measurements) or indirectly by measuring inhibition of binding of [³H]leucine (K$_I$ determinations). For K$_D$ determinations, 60-70 μg diluted binding protein preparation was dialyzed overnight against varying concentrations of radiolabeled ligand at constant specific activity (generally 50,000 cpm/nmol) in 10 mM potassium phosphate, pH 6.9.

For 5'5'5'-trifluoroleucine (TFL) K$_I$ determinations, binding across a range of radiolabeled ligand concentrations was measured in the presence of different concentrations of unlabeled D,L-TFL. K$_I$ values of the L isomer were derived graphically using the method of Stinson and Holbrook (15), with the assumption that the D isomer does not inhibit leucine binding. K$_I$ values were also determined by the method of Dixon (16): binding of a constant concentration of radiolabeled ligand was measured at several inhibitor concentrations. For percent inhibition experiments, binding of 0.5 μM [³H]leucine was measured in the presence and absence of 100-fold excess (50 μM) unlabeled inhibitor.

RESULTS

Solution of the X-ray diffraction pattern of LIV-BP with a leucine molecule associated with the unbound crystals suggested three residues that might account for the divergent specificity of LIV-BP and LS-BP based on their proximity to the leucine binding site and the fact that they are divergent in LS-BP (8). Changes in these residues should lead to alterations in ligand preference of each protein. To test this hypothesis, site-directed mutagenesis was used to change each of the three diverged residues in the putative leucine binding site to the corresponding residue in the other binding protein. In a second series of mutagenesis experiments, a common StyI restriction enzyme cleavage site was introduced at equivalent positions in the *livJ* and *livK* genes. This common restriction site permitted the construction of hybrid binding proteins consisting of the amino-terminal 124 residues of one binding protein followed by the remaining carboxyl-terminal residues of the other binding protein. The first 124 amino acid residues comprise about two-thirds of the N domain of each protein. These mutations are summarized in Table 2.

Each binding protein gene mutation obtained by site-directed mutagenesis or *in vitro* gene fusion was cloned downstream from the bacteriophage T7 promoter in the expression plasmid pT7-5 (17) and transformed into the T7 RNA polymerase strain BL21(DE3) (18). High level expression and purification of the binding proteins

Table 2. <u>Summary of binding protein mutations</u>

Protein	Residue	Change	Mutant Designation
LIV-BP	124	Silent StyI site	JSty
JSty	18	Tyr→Trp	JYW18
JSty	100	Ala→Gly	JAG100
JSty	276	Phe→Tyr	JFY276
LS-BP	124	Silent StyI site	KSty
KSty	18	Trp→Tyr	KWY18
KSty	100	Gly→Ala	KGA100
KSty	276	Tyr→Phe	KYF276
JSty 1-124 + KSty 125-346		Hybrid protein	JK
KSty 1-124 + JSty 125-344		Hybrid protein	KJ

was obtained as described in Experimental Procedures. Leucine binding protein represented between 70% and 90% of the shock fluid as judged by examination of Coomassie-stained gels. Protein yields were generally 7-10 mg total shock fluid protein from 100 ml of bacterial culture, which was enough protein for over one hundred individual binding assays. The KWY18 mutation was constructed in three independent site-directed mutagenesis experiments. However, in each case, no expression was observed when the pT7-5 construct was induced with IPTG.

The dissociation constants (K_D) for leucine, isoleucine, and valine were measured for JSty, KSty, and each of the mutant binding proteins. Binding of the tritiated ligand was measured over a range of ligand concentration that spanned the K_D for each protein. Calculation of the K_D was carried out as described by Scatchard (19) from a plot of nmol ligand bound divided by ligand concentration versus nmol ligand bound. The slope of the Scatchard plot is equal to $-1/K_D$. Several of the site-directed mutants displayed increased K_D values relative to wild-type, indicating reduced binding. For leucine, JYW18, JK, and KYF276 showed the largest affect; for isoleucine and valine, JYW18 and JK showed the largest increase in K_D. The KJ hybrid protein did not bind any of the labelled ligands. The K_D and K_I values obtained from all the binding studies are summarized in Table 3A.

5'5'5'-Trifluoroleucine (TFL) binds LS-BP with an affinity equal to that of leucine, but does not bind LIV-BP (3). TFL inhibition of leucine binding was measured for wild-type and mutant proteins two ways. First, the method of Stinson and Holbrook (15) was used to calculate K_I based on the effect of increasing TFL concentration on the K_D of leucine. Second, Dixon plots (16) were used to calculate the K_I based on a plot of the reciprocal of bound ligand versus TFL concentration at a constant [^{3}H]leucine concentration for KSty, KGA100, KYF276, and several JSty mutants. K_I values obtained by these two graphical methods were in good agreement with one another. Both KGA100 and KYF276 display reduced affinity for TFL to approximately the same degree. The reduction in TFL binding by KYF276, however, is proportional to the reduction in leucine binding, while for KGA100, TFL binding is specifically reduced. Interestingly, leucine binding to JAG100 is inhibited reasonably well by TFL. JSty, JK, and JFY276 are not inhibited by 100-fold excess of TFL.

Dixon plots (1/(nmol bound) vs [I]) were also used to determine the K_I values for isoleucine and valine inhibition of leucine binding to JYW18 and JK. Dixon plots were particularly useful when direct determination of K_D would have required much higher ligand specific activity than was practical. K_I values derived from the ordinate intercept of Dixon plots are not in complete agreement with K_D values from Scatchard plots (see Table 3, especially JYW18 and JK isoleucine binding constants). This is likely due to a combination of factors including the difficulty of

Table 3. Binding characteristics of leucine-binding proteins

A. Binding protein K_D and K_I values[a,b]

Protein	Leu	Ile	Val	TFL[c]
JSty	0.40±0.1	0.40	0.70±0.1	*
JYW18	8.0	8±1 (K_I) 13±2 (K_D)	14(K_I) 10.5±1(K_D)	59
JAG100	0.45	0.39	1.1	12.5±0.5
JFY276	0.80±0.13	0.36	1.7	*

Protein	Leu	Ile	Val	TFL
KSty	0.40±0.10	‡	‡	0.43±.03
KWY18				
KGA100	0.48±0.05	‡	‡	2.0±0.1
KYF276	1.5±0.3	‡0	‡	2.1±0.4

Protein	Leu	Ile	Val	TFL
KJ	*	*	*	N.D.[d]
JK	8.0±0.5	32 (K_I) 14.9 (K_D)	‡	*

B. Fold difference from wild-type activity[e]

Protein	Leu	Ile	Val	TFL
JSty	1	1	1	--
JYW18	20	20-30	15-20	N.A.[f]
JAG100	1	1	1.6	N.A.[f]
JFY276	2	1	2.4	--
KSty	1	--	--	1
KWY18				
KGA100	1.2	--	--	4.7
KYF276	4	--	--	4.9
KJ	∞	∞	∞	N.D.[d]
JK	20	40-80	∞	--

[a] K_D and K_I values are μM. Where ranges are given, the value is the mean ± the range for three or more independent measurements. All numbers are K_D's, except for TFL and as indicated.

[b] Symbols used: * indicates >>50 μM; ‡ indicates >>1 mM

[c] TFL - 5′5′5′-Trifluoroleucine; values represent L component of D,L mixture

[d] Not Determined

[e] Fold difference is mutant K_D/wild-type K_D or mutant K_I/wild-type K_I

[f] Not Applicable

accurately measuring the K_D from the Scatchard plot and small inaccuracies in the K_D for leucine that is used in calculation of the K_I.

In order to further explore the structural requirements of the leucine binding site, the two wild-type binding proteins JSty and KSty were tested against a wide variety of potential inhibitors. Binding of 0.5μM [³H]leucine was measured in the presence and absence of 50μM (100-fold excess) unlabeled inhibitor. The K_D for both proteins is 0.4μM. The results of these inhibition experiments are shown in Table 4, where binding in the presence of an inhibitor is expressed as percent of uninhibited binding. These results largely support previous findings (3, 6). *allo*-Isoleucine, isoleucine, threonine, and valine are the most effective inhibitors of JSty. TFL is the only analog with a similar degree of inhibition of KSty. Other analogs are known to be transported by the LIV-I/LS systems and exhibit moderate inhibition: alanine, BCH, and leucylglycine inhibit binding to JSty and azaleucine, norleucine, and leucylglycine inhibit binding to KSty.

These results suggest that the JSty protein is able to accommodate a branch at the ß carbon while KSty is not. Interestingly, *allo*-isoleucine is an effective inhibitor of JSty but *allo*-threonine is not. Thus, the ß-carbon branch is not sufficient to allow binding; apparently a polar characteristic of the binding pocket can discriminate between threonine and *allo*-threonine. This same characteristic of the JSty binding pocket might be expected to distinguish between leucine and TFL. The JAG100 mutation may permit alignment of the hydrophillic trifluoro group with the polar characteristic of the binding pocket. Analogs with a planar geometry at the γ-carbon (methylallylglycine), a branch at the α-carbon (cycloleucine) or that have a shorter (alanine and serine) or longer (norleucine) side-chain length are not effective inhibitors of either wild-type protein.

Several mutants derived from JSty were tested against four structurally similar analogs: isoleucine, *allo*-isoleucine, threonine, and *allo*-threonine (all of the L configuration). The *allo* designation indicates that the compound is an enantiomer of the L-isomer at the ß carbon. Essentially, the branch is oriented in the opposite direction in the L- and L-*allo* isomers. JSty exhibits nearly equal recognition of isoleucine and *allo*-isoleucine and distinguishes nearly absolutely between threonine and *allo*-threonine (Figure 2). JAG100 has lost some of the recognition of *allo*-isoleucine, but is otherwise nearly identical to JSty. JYW18, JFY276, and JK display reduced inhibition by threonine and increased inhibition by *allo*-threonine. JK also shows a reduced affinity for *allo*-isoleucine.

DISCUSSION

Eight mutations have been constructed in the putative leucine binding sites of LIV-BP and LS-BP. As can be seen in Figure 1 and Table 2, the position 18 and 276 mutations consist of exchanges among the bulky hydrophobic residues phenylalanine, tyrosine, and tryptophan. The position 100 mutations alter the presence of a single methyl group by a glycine/alanine exchange. Each of these point

Table 4. Inhibition by leucine analogs

Inhibitor	% of uninhibited binding	
	JSty	KSty
L-Alanine	78	100
L-*allo*-Isoleucine	11	100
L-*allo*-Threonine	114	105
ß-2-aminobicycloheptane carboxylic acid (BCH)	83	93
D,L-Azaleucine	95	82
Cycloleucine	96	100
L-Isoleucine	1.1	101
D-Leucine	97	91
D,L-Leucylglycine	74	61
Methylallylglycine	104	103
L-Norleucine	90	72
L-Serine	100	97
L-Threonine	16	109
D,L-Trifluoroleucine	86	3.5
L-Valine	7.5	98

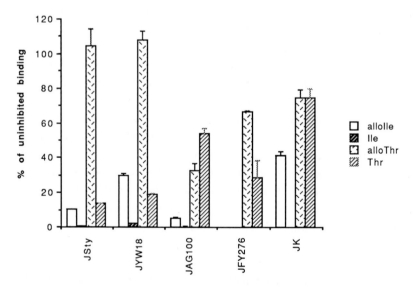

Figure 2. Inhibition of leucine binding to wild-type and mutant LIV-BP proteins by L- and L-*allo*- isoleucine and threonine. Binding of [³H]leucine was measured in the presence and absence of 100-fold excess of unlabeled inhibitor and plotted as the per cent of uninhibited binding. Values given are mean ± range for two experiments.

mutations resulted in alterations of the binding constants for one or more ligands relative to the wild-type proteins. The fold-reduction in binding activity compared to wild-type is presented in Table 3B.

The position 18 mutant, JYW18, bound leucine, isoleucine, and valine about twenty fold less well than JSty. Alteration of binding of each ligand by approximately the same amount suggests that this residue may participate in stabilization of the 'bound' or 'closed' conformation of the protein, rather than participating in leucine binding directly.

JFY276 showed a reduced affinity for leucine and valine by approximately two-fold, but bound isoleucine with a K_D indistinguishable from JSty. The other position 276 mutant, KYF276, showed an approximately four-fold reduction in leucine binding. TFL binding to KYF276 is reduced about five-fold, so TFL binding is reduced proportionately to leucine binding.

At position 100, both KGA100 and JAG100 showed leucine, isoleucine, and valine binding constants very similar to wild-type. TFL binding to KGA100, however, is reduced four-fold while leucine binding remains unchanged, indicating specificity for TFL is lost. Surprisingly, the corollary mutation, JAG100, exhibits a very reasonable K_I for TFL of 23 μM compared to wild-type LIV-BP and other 'J' mutants, which do not show any inhibition by TFL.

The hybrid binding proteins also display interesting substrate profiles. The KJ fusion did not bind any of the three tritiated substrates that were available — leucine, isoleucine, or valine. The JK fusion, on the other hand, bound leucine with a K_D of 8 μM, only about twenty times lower binding than the wild-type parent molecules. Isoleucine binding by JK was about 100-fold lower than wild-type, and valine binding was undetectable. This is the only observation of a discrimination in binding between the two ß-carbon-branched substrates isoleucine and valine.

As shown in Figure 2, the hydrophobicity of the substituents at the ß carbon contributes to the recognition of a ligand by LIV-BP. The ethyl group branch of isoleucine is reasonably well accommodated in either enantiomeric configuration, except in JK, where *allo*-isoleucine binding is reduced (Table 4). On the other hand, the hydroxyl group branch of threonine is not accommodated in the *allo* enantiomeric configuration by JSty. Disruption of the binding site in the JAG100, JFY276, and JK mutants causes a switch in preference toward *allo*-threonine from threonine. The JAG100 mutation gains specificity for both TFL and *allo*-threonine compared to wild-type, but the Ala->Gly change would not be expected to change the polarity of the binding pocket considerably.

Taken together, these results suggest that residues 18, 100, and 276 do participate in leucine binding. Several of the mutants show altered activity toward only one of the potential ligands, indicating a direct interaction with the ligand rather than a less well defined conformational alteration. Since none of the pairs of site-directed mutations described here resulted in an absolute switch of specificity for a particular ligand, other residues besides those suggested by Sack, *et al.* (8) are likely to contribute to binding of branched-chain ligands.

A complete description of ligand binding is also likely to require residues from the C domain of the protein in addition to the residues from the N domain that were examined in this study. The failure of the hybrid protein KJ to bind any of the branched-chain amino acids suggests that the intramolecular contacts necessary to cause the conformational changes necessary to close the binding cleft upon leucine binding might be missing. The JK protein must maintain sufficient cross-domain contacts to stabilize the 'closed' (liganded) conformation. Full understanding of the mechanism of differentiation among these very similar hydrophobic side chains will depend on the solution of X-ray crystal structures of the liganded forms of the binding proteins and some of the mutants described here.

REFERENCES
1. Quiocho, F.A., Sack, J.S., and Vyas, N.K. (1987) *Nature* **329**, 561
2. Adams, M.D., Wagner, L.M., Graddis, T.J., Landick, R., Antonucci, T.K, Gibson, A.L., and Oxender, D.L. (1990) *J. Biol. Chem.*, **265**, 11436
3. Rahmanian, M., Claus, D.R., and Oxender, D.L. (1973) *J. Bacteriol.* **116**, 1258
4. Harrison, L.I., Christensen, H.N., Handlogten, M.E., Oxender, D.L., and Quay, S.C. (1975) *J. Bacteriol.* **122**, 957
5. Christensen, H.N., Handlogten, M.E., Lam, I., Tager, H.S., and Zand, R. (1969) *J. Biol. Chem.* **244**, 1510
6. Penrose, W.R., Nichoalds, G.E., Piperno, J.R., and Oxender, D.L. (1968) *J. Biol. Chem.* **243**, 5921
7. Anderson, J.J. and Oxender, D.L. (1977) *J. Bacteriol.* **130**, 384
8. Sack, J.S., Saper, M.A., and Quiocho, F.A. (1989) *J. Mol. Biol.* **206**
9. Sack, J.S., Trakhanov, D.S., Tsigannik, I.H., and Quiocho, F.A. (1989) *J. Mol. Biol.* **206**, 193
10. Kunkel, T.A. (1985) *Proc. Natl. Acad. Sci.* **82**, 488
11. Su, T.-Z. and El-Gewely, M.R. (1988) *Gene* **69**, 81
12. Luria, S.E. and Burrows, J.W. (1957) *J. Bacteriol.* **74**, 461
13. Neu, H.C. and Heppel, L.A. (1965) *J. Biol. Chem.* **240**, 3685
14. Lowry, O.H., Rosebrough, N.J., Farr, A.L., and Randall, R.J. (1951) *J. Biol. Chem.* **193**, 265
15. Stinson, R.A. and Holbrook, J.J. (1973) *Biochem. J.* **131**, 719
16. Dixon, M. (1953) *Biochem. J.* **55**, 170
17. Tabor, S. and Richardson, C.C. (1985) *Proc. Natl. Acad. Sci. USA* **82**, 1074
18. Studier, F.W. and Moffatt, B.A. (1986) *J. Mol. Biol.* **189**, 113
19. Scatchard, G. (1949) *Annals N.Y. Acad. Sci.* **51**, 660
20. Piperno, J.R. and Oxender, D.L. (1968) *J. Biol. Chem.* **243**, 5914
21. Oxender, D.L., Anderson, J.J., Daniels, C.J., Landick, R., Gunsalus, R.P., Zurawski, G., and Yanofsky, C. (1980) *Proc. Natl. Acad. Sci. U.S.A.* **77**, 2005
22. Landick, R.C. (1984) *University of Michigan*, doctoral dissertation
23. Graddis, T.J. (1990) *University of Michigan*, doctoral dissertation

© 1991 Elsevier Science Publishers B.V. (Biomedical Division)
Site-Directed Mutagenesis and Protein Engineering
M.R. El-Gewely, editor.

THE MECHANISM OF PANCREATIC PHOSPHOLIPASE A$_2$ STUDIED WITH RECOMBINANT DNA TECHNIQUES

H.M. VERHEIJ

Department of Biochemistry , University of Utrecht, CBLE, University Center De Uithof, P.O. Box 80054, 3508 TB Utrecht (The Netherlands).

INTRODUCTION

Phospholipases A$_2$ (PLA2; E.C. 3.1.1.4) are enzymes that catalyze the stereospecific hydrolysis of the 2-*sn*-acyl ester bond of 3-*sn*-phosphoglycerides (1). More than 60 PLA2s have been isolated and sequenced from a wide range of sources including numerous snake and insect venoms and several cellular and extracellular sites in a variety of vertebrate species (2-3). Each of these enzymes exhibit a common set of structural and functional properties indicative of a family of highly conserved enzymes. All are relatively small (ca. 14 kDa), extensively (disulfide) cross-linked proteins. Even the more distant relatives display closely similar secondary and tertiary structures and a highly conserved catalytic machinery, Ca^{2+}-binding region and other structural attributes that apparently define the essential features required for the Ca^{2+}-dependent hydrolysis of aggregated phospholipid substrates that is characteristic of this family of enzymes (3-6).

Beyond this conserved framework, however, there can be substantial structural and functional differences. In particular, surface-exposed residues are highly variable giving rise to vast differences in the surface properties of these proteins, including charge (pI's range from ca. 4-11) and the propensity of a given PLA2 to form complexes in solution with either itself or other (regulatory) proteins (3). Two major structural determinants distinguish the pancreatic PLA2s from all other PLA2s. The pancreatic PLA2s are produced and stored in the pancreas in the form of an inactive precursor (Figure 1). Full activity requires tryptic activation in the intestine (7). No precursor forms have been reported for the venom and cellular PLA2s. In addition, all pancreatic PLA2s contain a surface loop (residues 62-72) of variable structure and in all other PLA2s this loop is lacking due to a 5-7 amino acid deletion. Based on the disulfide pairings the PLA2s can be devided into group I and II enzymes (8). The first group comprises elapid venom and pancreatic PLA2s, to the group II enzymes belong the extracellular enzymes from (pit) viper venoms and the cellular enzymes.

Chemical modification and X-ray crystallography studies (9-11) have indicated the importance of His-48, Asp-99 and Asp-49, the latter residue as a Ca^{2+}-binding ligand, for the enzymatic activity of pancreatic PLA2. Based on these data a general catalytic mechanism has been proposed (9), in which the aspartate-histidine couple and a water molecule serve as a catalytic triad (Figure 2), suggesting a similarity to the proton relay system of the serine esterases (12). According to the 3-D structure, the Asp-99 residue not only interacts with His-48, but forms hydrogen bonds in the interior of the enzyme as well (13). The Oδ1 atom can accept hydrogen bonds from Tyr-73 and from a water molecule, which in turn is H-bonded to the NH^{3+}-group of Ala-1 and to Tyr-52. The Oδ2 atom of

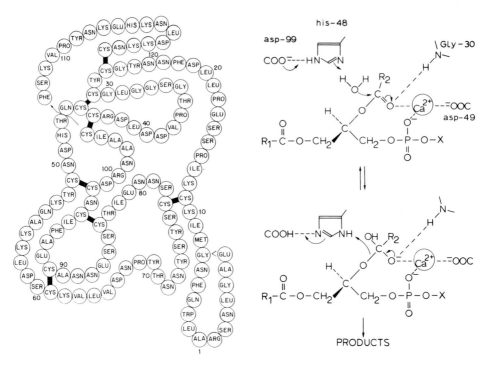

Figure 1. Primary structure of porcine pancreatic proPLA2. The tryptic activation to PLA2 occurs at the Arg[-1]-Ala[1] bond.

Figure 2. The proposed catalytic mechanism of PLA2.

Figure 3. The extended hydrogen bonding system of PLA2, showing the connection of His-48 to the amino terminal region.

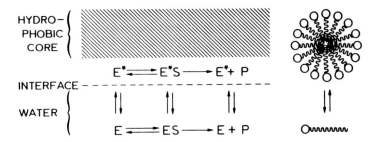

$$E^* \rightleftharpoons E^*S \longrightarrow E^* + P$$

INTERFACE

$$E \rightleftharpoons ES \longrightarrow E + P$$

HYDRO-PHOBIC CORE

WATER

Figure 4. Schematic representation of the interaction of PLA2 with monomeric substrates in the presence and in the absence of lipid-water interfaces.

Asp-99 can accept hydrogen bonds of the H atom of the N3 of His-48 and of Tyr-52 (Figure 3). The two tyrosines at position 52 and 73 are fully conserved in all seventynine known sequences of pancreatic-, snake venom- and intracellular mammalian phospholipases A2 (for a comparison of 50 sequences see 14).

During the last 25 years a wealth of information on the action of extracellular PLAs has been obtained using "classical" techniques, like chemical modification, semi-synthesis, protein sequencing and X-ray crystallography, several spectroscopic methods and kinetic assays with a large variety of synthetic substrates. These experiments have yielded the proposition of a catalytic mechanism both for the hydrolysis of monomeric substrates and for the interaction of the enzyme with aggregated lipids (9,15) that is still used as a working hypothesis (c.f. Figures 2 and 4). Without a crystal stucture of an enzyme-calcium- inhibitor complex it remains difficult to predict the orientation of substrate molecules in the active site. Hence the role of the cofactor calcium and of the water molecule which is presumed to be the nucleophile remain speculative. Even further away is an answer to the central question in lipolysis: what causes the efficient hydrolysis of aggregated substrates compared to that on monodisperse solutions? Recombinant DNA techniques have become a powerful tool in the study of proteins, permitting the replacement of any given residue by any other amino acid. The cloning and expression of porcine pancreatic PLA2 in *Escherichia coli* and yeast (16,17) allowing for the isolation of 50-100 mg quantities of (mutant) PLA2s, made it possible to verify the proposed mechanism of catalysis and to probe the residues involved in the interaction of calcium as well as with substrate molecules. In the present paper the results obtained by the combined use of mutant proteins, competitive inhibitors and X-ray crystallography will be discussed.

PHOSPHOLIPASE A2 MUTANTS

Natural variations in the PLA2 sequence

Despite the fact that the core sequence of all PLA2s is highly conserved, Nature has created a number of natural variants in which otherwise conserved residues have been changed. Examples of such variations are found in the calcium binding loop (Asp49→Lys and Gly30→Ser), and in a surface loop that is present in pancreatic PLA2s, but is lacking in all other PLA2s.

In order to probe the role of Asp-49 in the active site of PLA2, two mutant proteins were constructed containing either Glu or Lys at position 49. In addition a Lys-49 PLA2 was purified from the venom of *Agkistrodon piscivorus piscivorus* since this enzyme was reported (18) to have retained activity. The enzymatic activities and the affinities for substrate and for Ca^{2+} ions were examined. Conversion of Asp-49 to either Glu or Lys strongly reduces the binding of Ca^{2+} ions in particular for the lysine mutant but the affinity for substrate analogues is hardly affected. The results obtained both with the pancreatic PLA2 mutants and with the native venom enzymes show that Asp-49 as calcium ligand is essential for the catalytic action of PLA2.

The venom of the Australian tiger snake *Notechus scutatus* contains active PLA2s and an inactive PLA2 homolog that differs from the active species by at least 40 mutations (19). Among the substituted amino acids is the replacement of the otherwise conserved glycine-30 by a serine residue. In order to better understand the effect of the latter substitution, glycine-30 in porcine pancreatic PLA2 has been replaced by a serine (20). The resulting mutant G30S was expressed in *Escherichia coli*, purified and characterized. The mutation caused a significant drop in enzymatic activity towards monomeric and aggregated substrates, had a limited effect on substrate binding, whereas the affinity for calcium ions was reduced tenfold. The lowered catalytic efficiency may be caused by a slightly different position of calcium relative to His-48 and/or by a different orientation of the backbone amide NH of serine-30. Such a reorientation could result from a change in ϕ, ψ angles that can be anticipated upon substituting serine for glycine. Consequently the amide NH may no longer form a hydrogen bond with the ester carbonyl. Where in the proposed mechanism (9) this hydrogen bond together with the calcium ion stabilises the tetrahedral intermediate, the removal of this hydrogen bond could explain the observed reduced catalytic rate constants.

Protein engineering in combination with X-ray crystallography have been used to study the role of the surface loop that distinguishes pancreatic PLA2s from snake venom and cellular PLA2s (21). Removal of residues 62 to 66 from porcine pancreatic PLA2 does not change the binding constant for micelles signifcantly, but it improves catalytic activity up to 16 times on micellar (zwitterionic) lecithin substrates. In contrast, the decrease in activity on negatively charged substrates is greater than fourfold. A crystallographic study of the mutant enzyme shows that the region of the deletion has a well-defined structure that differs from the structure of the wild-type enzyme in the region of the deletion. No structural changes in the active site of the enzyme were detected. Recently a similar mutation has been described in *bovine* pancreatic PLA2 (22). In this case the activity on dioctanoyllecithin was lowered 2.5-fold in contrast to the 2.5-fold increase that we observed with the *porcine* pancreatic PLA2 mutant.

Surface residues near the entrance of the active site

Near the entrance of the active site of PLA2 two residues, *i.e.* 31 and 69, are present at the edge of the active site cleft and their side chains might be involved in the interaction with substrate molecules (see Figure 5). In all PLA2 sequences the residue at 69 is either Tyr or Lys, the residue at 31 is Leu, Trp, Arg or Ala in more than 85% of all PLA2s.

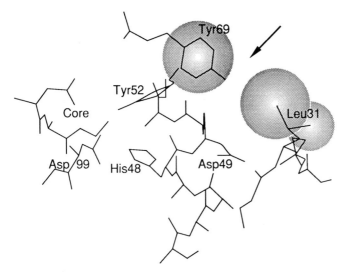

Figure 5. Cross section of the active site of PLA2 showing His-48, Asp-49 and Asp-99. The Van der Waals surfaces of Leu-31 and Tyr-69 are shaded. Substrate molecules are supposed to approach the active site from the direction of the arrow.

To study the role of residue 31 in porcine pancreatic PLA2, six mutant enzymes were produced by site-directed mutagenesis, replacing Leu by either Trp, Arg, Ala, Thr, Ser, or Gly (23). Direct binding studies indicated a 3 to 6 times greater affinity of the Trp-31 PLA2 for both monomeric and micellar substrate analogues, relative to the wild-type enzyme. The other five mutants posses an unchanged affinity for monomers of the product analogue n-decylphosphocholine, whereas the affinities for micelles of the monoacyl product analogue n-hexadecylphosphocholine were decreased 9-20 times. Kinetic studies with monomeric substrates showed that the mutants have Vmax values which range between 2 and 70% relative to the wild-type enzyme, the mutant with a Trp at position 31 being the most active enzyme. The results of these investigations point to a versatile role for the residue at position 31: involvement in the binding and orientating of monomeric substrate (analogues), involvement in the binding of the enzyme to micellar substrate analogues and possibly involvement in shielding the active site from excess water.

The role of Tyr-69 of porcine pancreatic PLA2 was studied with the help of proteins modified by site-directed mutagenesis in combination with the use of phospholipid analogues (24,25). Two mutants were used containing Phe and Lys, respectively, at position 69. Modifications in the phospholipids included among others a changed chirality at sn-2 of the glycerol backbone and the introduction of a sulfur at the phosphorus (thion phospholipids). Replacement of Tyr-69 by Lys reduces enzymatic activity, but the mutant enzyme retains both the stereo- and positional specificity of native PLA2. The Phe-69 mutant not only hydrolyses phospholipids of the natural L configuration, but the D-isomers are hydrolysed too. Similarly the Y69F mutant hydrolyses the R_p isomer of thion phospholipids more rapidly than normal phospholipids, but the S_p thion isomer is hydrolyzed too, although at a low (~4%) rate. Our data suggest that in PLA2 fixation of the phosphate group is achieved both by an interaction with the phenolic OH of Tyr-69 and by an interaction with the calcium ion. In the mutant Y69K the ε-NH$_2$ group can play a similar role as the Tyr-OH group in native PLA2. The Y69F mutant cannot bind and orient the phosphate group resulting in a reduced stereospecificity. Moreover the smaller sidechain of the Y69F mutant can interact with more bulky headgroups, allowing for relatively high enzymatic activities on modified phospholipids.

The extended hydrogen bonding system of PLA2

In the active center of all PLA2s His-48 is at hydrogen bonding distance to Asp-99. This Asp-His couple is assumed to act together with a water molecule as a catalytic triad. Asp-99 is also linked via an extended hydrogen bonding system to the side chains of Tyr-52 and Tyr-73. To probe the function of the fully conserved Asp-99, Tyr-52 and Tyr-73 residues in porcine PLA2, the Asp-99 residue was replaced by Asn, and each of the tyrosines were separately replaced by either a Phe, Leu or a Gln (26).

The catalytic and binding properties of the Phe-52 and Phe-73 mutants did not change significantly relative to the wild-type enzyme. This rules out the possibility that either one of the two Tyr residues in the wild-type enzyme can function as an acyl acceptor or proton donor in catalysis. The Gln-73 mutant could not be obtained in any significant amounts probably due to incorrect folding. The Gln-52, Leu-52 and Leu-73 mutants were isolated in low yields and had a 15-60 times decreased catalytic activity while their substrate binding was nearly unchanged. These results suggest a structural role rather than a catalytic function of Tyr-52 and Tyr-73: at this position an aromatic residue appears to be essential.

At position 99 substitution of asparagine for aspartate hardly affects the binding constants for both monomeric and micellar substrate analogues. Kinetic characterization revealed that the Asn-99 mutant has retained no less than 65% of its enzymatic activity on the monomeric substrate, while the Asp to Asn substitution decreases the catalytic activity on micellar substrates to about 4%. To explain these remaining activities we suggest that in the mutant the Asn-99 orients His-48 in the same way as Asp-99 orients His-48 in native PLA2. The lowered activity may be caused by a reduced stabilisation of the transition state because the protonated imidazole ring is now interacting with a neutral amide rather than with a negatively charged carboxylate. This reduction in rate is more pronounced for the hydrolysis of monomeric substrates than for the hydrolysis of aggregated substrates. In this respect, it must be kept in mind that aggregated substrates are hydrolysed at least three orders of magnitude faster than monomeric substrates and that proton transfer might be the rate limiting step for the hydrolysis of aggregated substrates but not for hydrolysis of monomeric substrate.

The three-dimensional structure of a PLA2-inhibitor complex.

Three-dimensional (3D) structures have been elucidated only for extracellular PLA2s (6,11,13) and it is evident that these structures are very similar (27). No structure is available for an intracellular PLA2, but based on the amino acid sequence homology the 3D-structures of the pancreatic PLA2-inhibitor complexes are probably suitable as a model for other extracellular and for intracellular

PLA2s. To obtain such a 3D-stucture, the competitive inhibitor (R) 2-dodecanoyl-amino-1-hexanol-phosphoglycerol was cocrystallized with a porcine PLA2 mutant which lacks residues 62-66 and which was selected because of its crystallisation properties. In order to increase the affinity for monomeric substrates still further, a Trp was introduced at position 31. The inhibitor binds about 1200 times stronger than the substrate (R) 1,2-didodecanoyl-glycerol-3-phosphocholine to this mutant PLA2. The structure of the complex of this mutant PLA2 and the inhibitor was solved and refined at 2.4 Å resolution (28). The inhibitor is bound to the calcium ion in the enzyme's active site both with its phosphate group as well as with the carbonyl oxygen of the amide bond (see Figure 6). These two ligands replace two water molecules that are ligands of the calcium ion in the native structure. The hydroxyl group of Tyr-69 makes a hydrogen bond with one of the oxygen atoms of the phosphate

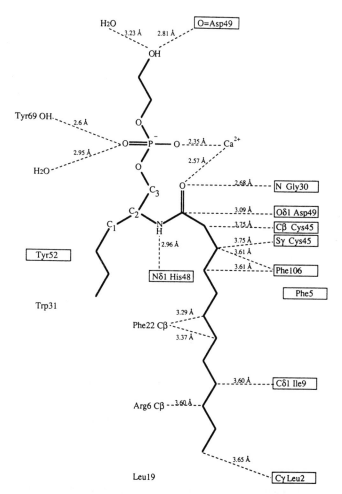

Figure 6. Schematic drawing of the interactions of the inhibitor with PLA2. Residues within 5 Å are drawn. Residues that are identical in the used mutant and in human cellular PLA2 (4) are boxed. The other residues are substitued as follows: R6H, L19A, F22Y, W31V and Y69K.

group. This is in agreement with site-directed mutagenesis experiments indicating that Tyr-69 is important for the precise positioning of the phospholipid substrate in the enzyme's active site (24,25). The acyl chain of the inhibitor makes extensive hydrophobic contacts with the disulfide bridge between residues 29 and 45, and the side chains of residues Leu-2, Phe-5, Ile-9, Leu-19. Phe-22 and Tyr-52. The nitrogen of the amide bond makes a strong hydrogen bond with the Nδ1 atom of His-48. The short alkyl chain has no interactions within 4 Å with the protein. The glycol headgroup makes a hydrogen bond to the backbone carbonyl oxygen of Asp-49. All residues involved in binding are either identical or substituted functionally in the human cellular PLA2.

The active site residues of PLA2 do not, within error, change in position upon binding of the inhibitor. However, two water molecules that are near the Nδ1 atom of His-48 in the native structure, have been replaced by the bound inhibitor. One of these water molecules was proposed to be activated by the His48-Asp99 couple and to act as the nucleophile attacking the carbonyl carbon atom of the scissile bond (Figure 2). The oxy-anion intermediate would then be stabilized by the calcium ion and the peptide NH group of Gly-30 (9,13). Indeed in the present structure the inhibitor's carbonyl oxygen atom is bound to the calcium ion and is at 2.7 Å from the NH group of Gly-30. However, a strong hydrogen bond exists between the NH of the inhibitor's amide bond and the Nδ1 atom of His-48. In a true substrate-enzyme complex the ester oxygen cannot form such a hydrogen bond with the unprotonated Nδ1 atom, which is the catalytically active species (9). This extra hydrogen bond can contribute to the higher affinity of the enzyme for the inhibitor compared with that for the natural substrate. The absence of a water molecule in the active site close to the acyl-amino linkage raises the question about the attacking nucleophile. Therefore, in a model building experiment, the amide bond of the inhibitor was replaced by an ester bond to see how a true substrate might bind in the active site and whether enough space would become available for a water molecule. Indeed, a simple tilting of the ester group away from His-48 creates sufficient space for such a water molecule. The modelled substrate is liganded to the calcium ion both by its phosphate group and the carbonyl oxygen atom. The modelled water is hydrogen bonded to the His-48 Nδ1 atom, and is at about 2.8 Å from the carbonyl carbon atom of the substrate.

SUMMARIZING COMMENTS

In the last few years a significant progress has been noticed on the way how the monomeric substrate (or inhibitor) molecule binds to the active site of PLA2. The recently determined X-ray structure agrees well with the early prediction of how a substrate molecule would fit into the active site. This structure is also in agreement with the conclusions that were based on the results of site-directed mutagenesis studies pertaining to the role of Gly-30, Leu-31, Asp-49 and Tyr-69 in fixating the substrate, the cofactor calcium and their respective roles in the true catalytic steps. Despite this progress in our knowledge on the precise positioning of the individual substrate molecule in the active site, the question how the enzyme is activated by interfaces is not solved yet. Further experiments on the extended hydrogen bonding system that links the amino-terminal helix to the true active site (Figure 3), in combination with NMR experiments might shed more light on this problem.

ACKNOWLEDGEMENTS

The author is indebted to many persons who contibuted in several ways to the work presented here. These contributions include the synthesis of oligonucleotides and phospholipids, site-directed mutagenesis, purification and characterisation of (mutant) PLA2s and many discussions. Part of this work has been supported by the EEC (BAP contract NL-0071) and by the Netherlands Foundation for the Advancement of Pure Science (NWO).

REFERENCES

1. de Haas, G.H., Postema, N.M., Nieuwenhuizen, W., and van Deenen, L.L.M. (1968) Biochim. Biophys. Acta 159, 103-117

2. Dufton, M.J., Eaker, D. and Hider, R.C. (1983) Eur. J. Biochem. 137, 537-544

3. Waite, M. (1987) in Handbook of Lipid Research (Hanahan, D.J. ed.), Vol. 5: The Phospholipases, 161-178, Plenum Press, New York

4. Kramer, R.M., Hession, C., Johansen, B., Hayes, G., McGray, P., Chow, E.P., Tizzard, R. and Pepinsky, R.B., 1989, J. Biol. Chem. 264, 5768-5775

5. Kuchler, K., Gmachl, W., Sippl, M.J. and Kreil, G. (1989) Eur. J. Biochem. 184, 249-254

6. Brunie, S., Bolin, J., Gewirth, D. and Sigler, P.B. (1985) J. Biol. Chem., 260, 9742-9749

7. de Haas, G.H., Postema, N.M., Nieuwenhuizen, W. and van Deenen, L.L.M. (1968) Biochim. Biophys. Acta 159, 118-129

8. Heinrikson, R.L., Krueger, E.T. and Keim, P.S. (1977) J. Biol. Chem. 252, 4913-4921

9. Verheij, H.M., Volwerk, J.J., Jansen, E.H.J.M., Puijk, W.C., Dijkstra, B.W., Drenth, J. and de Haas, G.H. (1980) Biochemistry 19, 743-750

10. Fleer, E.A.M., Verheij, H.M. and de Haas, G.H. (1981) Eur. J. Biochem. 113, 283-288

11. Dijkstra, B.W., Renetseder, R., Kalk, K.H., Hol, W.G.J.and Drenth, J. (1983) J. Mol. Biol. 168, 163-179

12. Stamato, F.M.L.G., Longo, E., Ferreira, R. and Tapia, O. (1986) J. Theor. Biol., 118, 45-59

13. Dijkstra, B.W., Kalk, K.H., Hol, W.G.J. and Drenth, J. (1981) J. Mol. Biol. 147, 97-123

14. van den Bergh, C.J., Slotboom, A.J., Verheij, H.M. and de Haas, G.H. (1989) J. Cellular Biochem. 39, 379-390

15. Pieterson, W.A., Vidal, J.C., Volwerk, J.J. and de Haas, G.H. (1974) Biochemistry 13, 1455-1461

16. de Geus, P., van den Bergh, C.J., Kuipers, O.P., Verheij, H.M., Hoekstra, W.P.M. and de Haas, G.H. (1987) Nucleic Acids Res. 9, 3743-3759

17. Bergh, C.J. van den, Bekkers, A.C.A.P.A., de Geus, P., Verheij, H.M. and de Haas, G.H. (1987) Eur. J. Biochem. 170, 241-246

18. Maraganore, J.M. and Heinrikson, R.L. (1986) J. Biol. Chem. 261, 4797-4804

19. Lind, P. and Eaker, D. (1980) Eur. J. Biochem. 111, 403-409

20. Bekkers, A.C.A.P.A., Franken, P.A., Toxopeus,E., Verheij, H.M. and de Haas, G.H., submitted for publication

21. Kuipers, O.P., Thunnissen, M.M.G.M., de Geus, P., Dijkstra, B.W., Drenth, J., Verheij, H.M., and de Haas, G.H. (1989) Science 244, 82-85

22. Kimura, S., Tanaka, T., Shimada, I., Shiratori, Y., Nakagawa, S., Nakamura, H., Inagaki, F. and Ota, Y. (1990) Agric. Biol. Chem. 54, 633-639

23. Kuipers, O.P., Vis, R., Kerver, J., Verheij, H.M. and de Haas, G.H. (1990) Protein Engineering 3, 599-603

24. Kuipers, O.P., Dijkman, R., Pals, C.E.G.M., Verheij, H.M., and de Haas, G.H. (1989) Protein Engineering 2 , 467-471

25. Kuipers, O.P., Dekker, N., Verheij, H.M. and de Haas, G.H. (1990) Biochemistry, 29 6094-6102

26. Kuipers, O.P., Franken, P.A., Hendriks, R., Verheij, H.M. and de Haas, G.H. submitted for publication

27. Renetseder, R., Dijkstra, B.W., Huizinga, K., Kalk, K.H. and Drenth, J. (1988) J. Mol. Biol. 200, 181-188.

28. Thunnissen, M.M.G.M., AB, E., Kalk, K.H., Drenth, J., Dijkstra, B.W., Kuipers, O.P., Dijkman, R., de Haas, G.H. and Verheij, H.M., submitted for publication

© 1991 Elsevier Science Publishers B.V. (Biomedical Division)
Site-Directed Mutagenesis and Protein Engineering
M.R. El-Gewely, editor.

PROCESSING AND STABILITY STUDIES OF RECOMBINANT HUMAN PARATHYROID HORMONE BY IN VITRO MUTAGENESIS

KAARE M. GAUTVIK*, B. NAJMA KAREEM*, SJUR REPPE*, ERIK ROKKONES*, OLE KR. OLSTAD*, ODD S. GABRIELSEN**, ANDERS HØGSET[+], OLA R. BLINGSMO*, VIGDIS T. GAUTVIK*, PETER ALESTRØM[++], and TORDIS B. ØYEN**

*Institute of Medical Biochemistry, University of Oslo, p.b. 1112 Blindern, 0317 Oslo 3
**Institute for Biochemistry, University of Oslo, p.b. 1041 Blindern, 0316 Oslo 3
[+]Nycomed, p.b. 4220 Torshov, 0401 Oslo 4
[++]Agricultural University of Norway, p.b. 36, N-1432 Ås-NLH, Norway.

INTRODUCTION

Secretory proteins are faced with two major problems after being translated. First, they need to be directed into the endoplasmic reticulum and guided through the secretory pathway. Secondly, they have to be released in an undegraded form. In order to solve these problems, both eukaryotic and prokaryotic secretory proteins are synthesized as larger precursors containing N-terminal extensions. These extensions aid the entering of the proteins into the endoplasmatic reticulum to the secretory pathway via the Golgi apparatus or secretory granules in eukaryotic cells. In prokaryotic cells they are removed, once export of the protein is under way, through the action of an endoproteolytic "signal peptidase". For the secretory protein it is of vital importance that only the normal processing sites are exposed and that other potential cleavage sites are hidden from undesired proteolysis. The primary amino acid sequence defines the processing sites and is also of import-

---

The abbreviations used are:

- human parathyroid hormone, hPTH
- PTH parathyroid hormone
- SDS, sodium dodecyl sulfate
- HPLC, high performance liquid chromatography

ance for the three-dimensional structure of a protein. However, today there is no molecular simulation alone that will allow us to define which of the multitude of conformers available to a polypeptide that will occur in the environments of a living organism. In contrast to studies that predict folding of a given protein based on its primary amino acid sequence as well as knowledge of the three-dimentional structure of a homologue, we have studied the importance of small primary changes in the prepro-peptide on cellular translocation and the overall stability of the same protein, namely human parathyroid hormone.

Human parathyroid hormone (hPTH) which is important for calcium homeostasis (1), is produced as a 115 amino acid prepropeptide where the first 25 amino acids represent the signal sequence. The propart of the hormone consists of 6 amino acid recidues giving rise to a mature hormone of 84 amino acid recidues (2-6). The function of parathyroid hormone is to act as a physiological bone builder and also to increase serum calcium by its action on the kidney.

The hormone is secreted from the parathyroid glands in response to a lowering of serum ionized calcium, and its half-life in blood is in the order of 30 min. (7).

The present studies have demonstrated the importance of species specific signal sequences for secretion of hPTH. Moreover, the N-terminal prepro extensions play an essential role in folding of the hormone and modulate the intracellular degradation in three host systems, mammalian cells, bacteria and yeast.

The results from our work in microorganisms have suggested that an especially sensitive internal processing site in hPTH is after lys in position 26, which may also represent its physiological cleavage site (8-10).

MATERIAL AND METHODS

Materials: Restriction enzmes and other DNA-metabolising enzymes were purchased from New England Biolabs. ^{125}I-Antibody -IgG was from

Amersham Corp., and NH_2-terminal-specific anti-PTH antibody was bought from CHEMICON. Synthetic hPTH (1-84) was from Sigma.

Bacterial strains and plasmids: E.coli strains DH5 and BJ5183 (obtained from Dr.F.Lacroute, Centre de Genetique Moleculaire du C.N.R.S., Gif-sur.Yvette, France) containing plasmids were described earlier (10).

Cell Lines and Culture Conditions

Mouse mammary tumour celline (C127I, ATCC CRL 1616) and Chinese hamster lung celline (DON ATCC CCL 16) were obtained from American Tissue Culture Collection (ATCC), USA. Both cell lines were cultured in monolayer in 37°C and 5% Co_2 in air. C127I cells were cultured in Dulbecco's Modified Eagle Medium and DON cells in McCoy's 5A Medium (GIBCO BRL, USA), both supplemented with 5% fetal calf serum (FCS) and penicillin 50 IU/ml and streptomycin 50 µg/ml (GIBCO BRL, USA). For large scale culturing of the cells 900 cm² roller bottles (Costar Europe Ltd.) were used, and the serum level was lowered to 1%.

Transfection of host cells and selection of clones

The expression vectors decribed above were transfected separately into both C127 cells and DOn cells using calcium phosphate precipitation method (11). The cells were co-transfected with the vector pKGE53 which contains the gene coding for neomycin resistance under control of Harvey Sarcoma 5'LTR.

Several clones were isolated after cultivation in G-418 sulphate (Geneticin, GIBCO, Scotland) containing medium (1.5 mg G-418/ml). The isolated clones were grown individually, and analysed for expression of hPTH.

Strains and culture conditions

The S. cerevisiae strains used were FL200 (α, leu2, ura3) and BJ1991 (α Trp1, ura3-52, leu2, prb1-112, pep4-3). Yeast cells were transformed by standard methods. Transformants were selected by plating on selectiv medium. Yeast cells were grown at 30°C in YNB medium (0.67% yeast nitrogen base, 2% glucose, supplemented with amino acids (50-75 µg/ml), and adenine and uracil (10 µg/ml) or YNB medium supplemented with 1% casamino acids (Difco).

DNA construction and sequencing: DNA methodology was performed
essentially according to Maniatis et al. (12). DNA sequencing was
performed on plasmid DNA with Sequenase (United States Biochemical
Corporation) in accordance with the suppliers manual.

Screening for the production of protein A: Immunological screening was used for the identification and isolation of clones producing Staphylococcus aureus protein A in conjuction with the human
parathyroid hormone as described (13).

Lysated and washed cells on the nitrocellulose filters were incubated with ^{125}I labelled IgG anti rabbit (0.3 uCi/ml) in Tris buffer
saline with 3% (w/v) bovine serum albumin, washed extensively in
Tris buffer saline containing 1% Tween and then autoradiographed.

Cultivation and Preparation of Cellular Fractions: The cultivation was performed in 2 l shaking flask containing 500 ml medium.
Cells were harvested by centrifugation at 5000 x g for 20 minutes,
and supernatant was taken as the growth medium fraction. The periplasm fraction was prepared by an osmotic shock method. The soluble
intracellular fraction was prepared by sonicating the cell pellet
remaining after extraction of the periplasmic fraction.

Protein purification: The fusion protien of hPTH and protein A
from the different cell extract was purified by IgG affinity chromatography using fast flow IgG - Sepharose (Pharmacia, Sweden).

Construction of plasmids for expression of α-factor hPTH fusions

A BglII EcoRI fragment from the entire pMFα1 gene was inserted
into yeast E.coli shuttle vector (9). The resulting expression
vector contains the α-factor promoter and transcription terminator
as well as the entire α-factor gene (9). To insert the parathyroid
hormone gene in frame, a DpnI SalI fragment from a hPTH/pUC19
subclone that encodes the mature hPTH sequence with its stop codon,
was cloned into the pαLX vector between a filled in HindIII site
and a SalI site.

The plasmid containing this expression cassette was prepared from
E.coli, and a BglII BclI fragment containing the whole fusion gene
with its promoter and terminator was excised and transfered to
yeast shuttle vectors (9).

In vitro mutagenesis

In order to delete the STE13 recognition sequence located immediately N-terminal to hPTH by site directed in vitro mutagenesis of the fusion gene, a 1495 bp XbaI fragment containing the α-factor promoter, the α-factor leader sequence and the hPTH gene including the stop codon, was isolated from pαLXPTH and subcloned into M13 mp19 (to give M13PTH-1). An oligonucleotide with the sequence GGATAAAAGATCTGTGAG was annealed to single stranded DNA prepared from the recombinant phage. The first ten nucleotides of the oligonucleotide are complementary to the sequence of the α-factor leader just preceding the Glu-Ala-Glu-Ala coding region, and the last eight nulceotides are complementary to the beginning of the hPTH sequence. After second strand synthesis and ligation, closed circular heteroduplex DNA was isolated after sedimentation through an alkaline sucrose gradient, and used to transform an BMH 71-18 mutL strain of E.coli defective in mismatch repair (kindly provided by Dr. G. Winter). Positive clones with the looped out sequence 3'-CTCCGACTCCGA-5' deleted, were identified by colony hybridization, using the mutagenizing oligonucleotide as the probe, and confirmed by DNA sequencing. The plasmid in these clones was designated M13PTH-2. The α-factor transcription terminator was then inserted into one of the positive M13 clones as a SalI HindIII fragment isolated from pMFα1 for transfering the entire expression cassette (9).

Immunological detection of hPTH in culture media

Radioimmunoassay of hPTH was carried out as described (7) using an antiserum specific for the 44-68 amino acid domain. Yeast culture media were assayed directly. Periplasmic fractions were obtained, after incubation of washed cells for 1 hour at 28°C in a buffer containing 0.05% Triton X-100, 0.1M NaH_2PO_4, and 0.5M NaCl. Intracellular fractions were prepared from the remaining cells by disruption in an Eaton press.

For electrophoretic analysis, yeast culture media were adjusted to pH 3 and concentrated by passage through a S-Sepharose Fast Flow column (Pharmacia AB), as described (9). The concentrated medium was analysed by 15% SDS/polyacrylamide gel electrophoresis (8), and either stained with silver (8,9) or further analysed by protein immunoblotting using Immobilion PVDF Transfer Membranes (Millipore) (9,10). Reference hPTH(1-84) was purchased from Peninsula Laboratories (USA). Protein blots were visualised with rooster anti-hPTH antiserum (7) as the primary antibody, rabbit anti-rooster IgG, and donkey [^{125}I]-antirabbit-IgG (Amersham) as secondary and tertiary antibody, respectively. Autoradiography was performed as described (7).

Purification of hPTH from yeast culture medium

The concentrated medium from the S-Sepharose column was subjected to further purification by reversed phase HPLC using a Vydac protein peptide C18 column (The Separation Group, Hesperia, CA, USA). The column was eluted with a linear gradient of acetonitrile/0.1% trifluoroacetic acid.

Amino acid sequence analysis

Proteins to be sequenced was purified either by acetic acid polyacrylamide gel electrophoresis and electroblotted directly onto glassfiber sheets (9) or by SDS polyacrylamide gel electrophoresis followed by blotting onto polyvinylidene difluoride membranes (10). Automated Edman degradation was performed on a 477A Protein Sequencer with an on-line 120A phenylthiohydantoin amino acid analyser from Applied Biosystems (Foster City, CA, USA). All reagents were obtained from Applied Biosystems (England).

RESULTS

Cloning and expression of human parathyroid hormone

cDNA was synthesized from poly(A) RNA by reverse transcription and cloned in plasmid pBR322 as described previously (8). Several full length clones were obtained. One of these is shown in Fig. 1.

```
TATGATGATACCTGCAAAAGACATGGCTAAAGTTATGATTGTCATGTTGGCAATTTGTTT
   MetIleProAlaLysAspMetAlaLysValMetIleValMetLeuAlaIleCysPh
  -25
```

```
        70              90             110
TCTTACAAAATCGGATGGGAAATCTGTTAAGAAGAGATCTGTGAGTGAAATACAGCTTAT
eLeuThrLysSerAspGlyLysSerValLysLysArgSerValSerGluIleGlnLeuMe
                        -6               +1
```

```
       130             150             170
GCATAACCTGGGAAAACATCTGAACTCGATGGAGAGAGTAGAATGGCTGCGTAAGAAGCT
tHisAsnLeuGlyLysHisLeuAsnSerMetGluArgValGluTrpLeuArgLysLysLe
```

```
       190             210             230
GCAGGATGTGCACAATTTTGTTGCCCTTGGAGCTCCTCTAGCTCCCAGAGATGCTGGTTC
uGlnAspValHisAsnPheValAlaLeuGlyAlaProLeuAlaProArgAspAlaGlySe
```

```
       250             270             290
CCAGAGGCCCCGAAAAAAGGAAGACAATGTCTTGGTTGAGAGCCATGAAAAAAGTCTTGG
rGlnArgProArgLysLysGluAspAsnValLeuValGluSerHisGluLysSerLeuGl
```

```
       310             330             350
AGAGGCAGACAAAGCTGATGTGAATGTATTAACTAAAGCTAAATCCCAGTGAAAATGAAA
yGluAlaAspLysAlaAspValAsnValLeuThrLysAlaLysSerGlnEnd
```

```
       370             390             410
ACAGATATTGTCAGAGTTCTGCTCTAGACAGTGTAGGGCAACAATACATGCTGCTAATTC
```

```
       430
AAAGCTCTATTA.
```

Fig. 1. Cloned DNA sequence coding for human preproparathyroid
hormone with flanking sequences. The corresponding amino acid
sequence of preproparathyroid hormone is shown. The signal sequence (pre-) part is stipled and the intact hormone is underlined.

The entire prepro gene was recloned in a mammalian expression
plasmid driven by the metallothioneine promoter for expression of
the intact hormone (Rokkones et al., unpublished) (Fig. 2).

The plasmids pSPTH37ZZ and pKP43PTH37ZZ were generated for expression in E.coli where the hPTH gene (111 base pairs) was placed
between the promoter and the signal sequence of Staphylococcus
Aureus protein A and the gene coding for two synthetic IgG binding
domains of the same protein (denoted ZZ) (see Fig. 3).

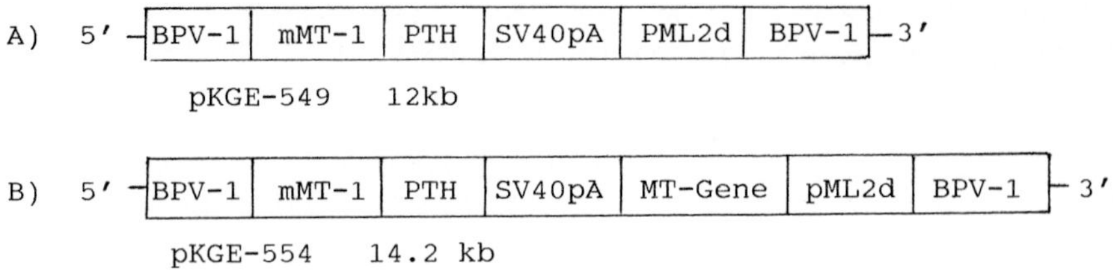

A)   5' ⊣| BPV-1 | mMT-1 | PTH | SV40pA | PML2d | BPV-1 |⊢ 3'

          pKGE-549   12kb

B)   5' ⊣| BPV-1 | mMT-1 | PTH | SV40pA | MT-Gene | pML2d | BPV-1 |⊢ 3'

          pKGE-554   14.2 kb

Fig. 2.   Diagram of the hPTH expression vectors in mammalian cells.
A) pKGE-549. B) pKGE-554. The directrion of transcription is unidi-
rectional 5'-3'. The mammalian cells used were from mouse mammary
tumor (C1271 cells) and Chinese hamster lung (DON cells). BPV-1,
Bovine papilloma virus type 1; mMT-1, murine metallothioneine 1
upstream regulatory element; MT-gene, murine metallothioneine gene;
PTH, human parathyroid hormone cDNA; SV40pA, simian virus 40 small
t-antigen intron and early polyadenylation signal; pML2d, a pBR322
derivative.

   These plasmids were constructed in order to study the effects of
systematically introduced changes in the bacterial signal sequence
on the trafficing and processing of the fusion protein between
protein A and the first 37 amino acid residues of hPTH. With this
strategy, a fusion protein consisting of a signal sequence of
protein A fused to a truncated hPTH gene, and IgG binding domains
of protein A should be translated. Both constructs should translocate the fusion protein into the secretory pathway of E.coli, but
only the first construct contains a correct cleavage site for the
E.coli processing enzyme (Fig. 3). The protein A ZZ part ensures
an efficient purification procedure using IgG-immunoaffinity procedures.

   For expression of hPTH gene in S.cerevisiae, a plasmid was constructed using the signal sequence of the mating factor α and its
transcription termination sequence (9). A modified signal sequence
was generated by in vitro mutagenesis (see Material and Methods)
where the processing site for the STE13 enzymes were deleted (9).
Fig. 4 A and B is a schematic presentation of the different gene
constructs used in these experiments for expression of hPTH in
S.cerevisiae.

Production and secretion of hPTH using the human signal sequence

Fig. 2 shows the gene construct for expression of hPTH in hamster and mouse cells.

After transfection of the cells (see Material and Methods), PTH was recovered from the medium. SDS polyacrylamide gel electrophoresis showed the presence of two forms of PTH as demonstrated by protein staining and immunoblots. The data revealed that the intact hormone and a peptide with molecular weight in the range between pro- and prepro-hormone were generated. No degradation forms were observed (Rokkones et al., unpublished).

When the human prepro sequence was used in E.coli or in S.cerevisiae, no secretory products were observed in the medium.

Expression of hPTH in E.coli using the natural (pSPTH37ZZ) and a mutagenized (pKP43PTH37ZZ).

The construction of the expression plasmid is described earlier (10) and its 12 bp deleted form is outlined in Fig. 3. In this construction the protein A promotor and signal sequence are used together with its termination signals. The construction is meant to give translation of a hybrid protein consisting of the first 37 amino acids of hPTH linked to IgG binding regions(the ZZ-domain) of protein A (Fig. 3). When this hybrid protein is correctly N-terminally cleaved, it shall result in a fusion protein of 19 kD.

In the pSPTH37ZZ, the intact signal sequence of protein A was used while in the mutated construct, pKP43PTH37ZZ, 12 bp were deleted from the signal sequence thereby removing -ALA-ALA-ASN-ALA which are situated adjacent to the last GLY in the signal sequence and specifying the cleavage site for the E.coli processing enzyme (Fig. 3, arrow). This construction should generate the hybrid protein, but the signal sequence cannot be cleaved. After transformation of the E.coli strain BJ 5183, the PTH related proteins were concentrated from medium, periplasm and intracellular compartment. These fractions were then passed through an IgG fast flow column. After washing, the hybrid protein was eluted, concentrated and fractions were run on gel electrophoresis. The hormone was characterized by immunoblotting, HPLC and finally N-terminal amino acid sequence analysis. Table 1 gives a summary of the results describing the PTH-related peptides in the cytoplasm, periplasm as well as those recovered from the medium using the two gene constructs. Only the construction with the normal signal sequence

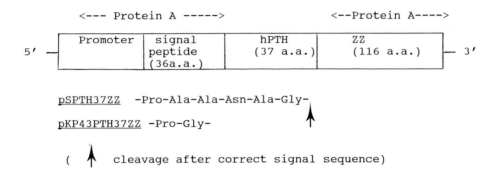

<--- Protein A ----->   <--Protein A---->

| Promoter | signal peptide (36a.a.) | hPTH (37 a.a.) | ZZ (116 a.a.) |

5' —   — 3'

pSPTH37ZZ -Pro-Ala-Ala-Asn-Ala-Gly-

pKP43PTH37ZZ -Pro-Gly-

(cleavage after correct signal sequence)

Fig.3. Expression plasmids (pSPTH37ZZ and pKP43PTH37ZZ) for production of hPTH-ZZ fusion protein.

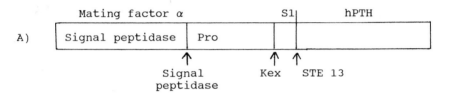

Mating factor α   S1   hPTH

A)

| Signal peptidase | Pro | | |

Signal   Kex   STE 13
peptidase

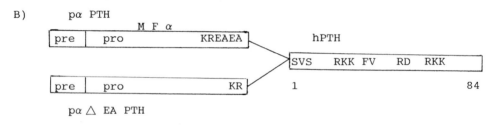

B)   pα PTH

M F α

| pre | pro | KREAEA | → | hPTH |

| SVS | RKK FV | RD | RKK |

| pre | pro | KR | → | 1 | 84 |

pα △ EA PTH

Fig. 4. A scheme presenting two gene constructs for expression of intact hPTH using the mating factor α system. A) The signal peptide (the prepart) is normally cleaved by a specific signal peptidase. The proregion is later processed by trypsin-like enzymes belonging to the KEX2 group cleaving after dibasic amino acids and the mating factor α is then released by processing of the (Glu-Ala) dipeptide which is located between the α factor spacer sequence (S1) and the first amino acid of hPTH. B) The plasmid pαPTH is constructed to translate the fusion peptide described in part A, and the arrows indicate dibasic amino acids in the mating factor α pro-region as well as possible processing sites within hPTH. In addition, the first N-terminal amino acids in the hormone are indicated. In the lower construct (pα△ EA PTH) 12 bases representing the EAEA sequence have been deleted so that this fusion protein is no longer dependent upon the action of the dipeptidyl amino peptidase STE 13 enzyme system, but would be processed to give a correct N-terminal sequence of the hormone by the action of KEX2 alone.

pSPTH37ZZ is able to show the appearance of a correctly cleaved
hPTH37ZZ protein of 19 KD which is present in all three compart-
ments. This construction is also the only one that results in
degradation products. One cleavage site was within the PTH part
after Lys residue 26 which belongs to a tribasic amino acid sequence (Fig 1). It is noteworthy that the major part of PTH related peptides is present in the medium and periplasm as degradation products using plasmid pSPTH37ZZ with the correct signal
sequence. The converse is the case when plasmid pKP43PTH37ZZ was
used. Almost no degradation products were detected and higher
molecular forms prevailed.

This result suggest that the major part of the degradation takes
place in conjunction with passing of the internal and external
membrane or in the periplasmic opace. The deleted signal sequence
in plasmid pKP43PTH37ZZ gives rise to a completely different profile of produced hPTH- ZZ hybrid proteins. As expected, no intact
and correctly cleaved hybrid protein was observed as indicated with
the negative demonstration of a peptide of 19 kD (Table 1). On the
other hand, hPTH-ZZ was represented as higher molecular weight
proteins of size equal to or larger than 22 kD corresponding to the
hybrid protein containing the uncleaved signal sequence. However,
this modified signal sequence is able to translocate effectively
the fusion protein into the medium and the periplasmic space.

Testing of the importance of processing sites in the prepro region
of the mating factor α in relation to expression of intact hPTH in
S.cerevisiae

The mating pheromone α-factor expression system has been used to
successfully express and export heterologous proteins from yeast
(14-17). The usual concept has been to express the proteins fused
to the prepro region of the mating factor α. During the secretion
process, the α factor leader sequence is cleaved off by an endopep-
tidase specific for a dibasic amino acid sequence encoded by the
KEX2 gene (18,19). This strategy works in some gene constructs to
liberate the mature heterologous protein. In other gene constructs
the KEX2 cleavage result in an N-terminal extension of the α-factor
spacer peptide which is subsequently cleaved off by a dipeptidyl
amino peptidase encoded by the STE13 gene (18,19).

Finally, the mature heterologous protein is then secreted into
the medium. The α-factor processing pathway itself has many similarities with the processing of precursor proteins like preprohPTH

Table 1.
Effects of mutagenesis in the signal sequence on cellular localization and processing of hPTH37ZZ fusion protein in E.coli.

Plasmid	Kilo Dalton		
	$< 16^{++}$	19^{*}	$< 22^{+}$
A. Medium			
1. pSPTH37ZZ	+++	+	−
2. pKP43PTH37ZZ	+	−	+++
B. Periplasm			
1. pSPTH37ZZ	++(+)	+	−
2. pKP43PH37ZZ	+	−	+++
C. Cytoplasm			
1. pSPTH37ZZ	+	++	−
2. pKP43PTH37ZZ	+	−	+

*Correctly cleaved and intact hPTH
+ Incompletely cleaved with N-terminal extension
++ Degraded forms

in mammalian cells (20,21). The original cDNA clone for human
parathyroid hormone was employed to make a gene fusion between the
yeast α-factor leader sequence where the first codon of the mature
hPTH sequence followed the last codon of the α-factor spacer sequence. Construction of the plasmid is briefly described in Material and Methods (9). To study the influence of the KEX2 and
STE13 enzyme systems on the cellular location and processing of
hPTH, two constructs were made (Fig. 4A and B). The first one is
the construct described above while the other represents a 12 base
pair deletion removing the Glu-Ala-Glu-Ala sequence (Fig. 4B) by in
vitro mutagenesis (see Material and Methods). The first constructs
then contained the dipeptidyl amino peptidase, STE 13, cleavage
site specified by the Glu-Ala sequence, and the other contained
Lys-Arg specifying the cleavage site for the KEX2 processing enzyme. Thus, for correct processing of the hPTH N-terminal, the
first construct required the coordinate participation of two processing enzyme systems while in the latter situation only the
protease-like KEX2 enzyme was required. After transformation of

S.cerevisiae, the cells were grown to late exponential phase and
medium and cells harvested. PTH from medium, intracellular space
and cytoplasm were purified, and quantitated (9). Table 2 summarizes the results and describe the effects of the Glu-Ala dipeptide
deletion on the cellular localization and processing of the intact
hormone. The construction containing the entire prepro region
including the Glu-Ala dipeptide sequence (PαPTH) gives rise to
correct N-terminal processing of the hormone, but also Glu-Ala
extended forms were discovered suggested an incomplete action of
the dipeptidyl peptidase (STE 13). The major part of the hormone
was found in the medium as fragments and the cleavage site was
localized to position 26-27 where Arg-Lys-Lys is present in the
hormone sequence. The other plasmid construct
PαΔ EA PTH, where the translated product required only KEX enzyme
processing, gave rise to a higher yield of the intact hormone, and
removed the presence of Glu-Ala extended peptides. On the other
hand, the amount of degraded hormone was not significantly altered.

   Substantial amount of hormone were only found in the medium
suggesting that the α-factor prepro part functions as a very effective translocator of this heterologous peptide into the secretory
pathway for extracellular discharge.

DISCUSSION AND CONCLUSIONS

   Even if there are similarities between the signal sequence in
mammals and in prokaryotic and eukaryotic systems, the human prepro
peptide sequence did not work in E.coli and in yeast as secretory
signals as opposed to the results from expression in mammalian
cells.

   Using species specific signal sequences of E.coli and
S.cerevisiae, the combined results showed that not only was the
hormone directed to the extracellular space, but the upstream
sequences conveyed important folding characteristics to the peptide.

   Both in E.coli and in S. Cerevisiae it has become clear that
changes in amino acid sequences adjacent to the N-terminal processing site of human PTH, give significant alterations in intracellu-

Table 2
Effects of mutagenesis in the mating factor α for cellular localization and processing of hPTH in S.cerevisiae.

Plamsid	Kilo Dalton		
	$<7.5^{++}$	9.5^{*}	$>12.5^{+}$
A. Medium			
1. pαPTH	+++	+	++
2. pαΔ EAPTH	++	++(+)	++
B. Periplasm			
1. pαPTH	(+)	(+)	(+)
2. pαΔ EAPTH	(+)	(+)	(+)
C. Cytoplasm			
1. pαPTH	(+)	(+)	(+)
2. pαΔ EAPTH	(+)	(+)	(+)

*Correctly cleaved and intact hPTH
+ Incompletely cleaved with N-terminal extension
++ Degraded forms

lar location as well as in the efficiency of correct processing. Also to our surprise, the signal sequence of protein A seems both to be very efficient in directing a translocation of the hybrid hPTH-ZZ protein not only into the periplasmic space, but all the way to the medium, and that an uncleavable signal sequence in some way protects PTH from being degraded by internal proteolysis.

In yeast, the two major processing enzymes, representing the KEX 2 and the STE 13 system, are able to perform a correct N-terminal processing of human PTH, but the latter enzyme system has a reduced capacity in relation to the KEX enzyme in the yeast strain used.

Both expression plasmids in yeast gave a significant amount of degraded hormone where a major cleavage site was located after the lysine residue in position 26, while internal potential basic sequences were used only to a minor extent. This indicates that the latter potential processing sites were probably shielded from attack of proteases due to the three dimentional folding of the peptide.

From these studies, the following conclusions can be made:

1. The human pre-pro sequences of parathyroid hormone are not functional as secretory signals in E.coli or in yeast.

2. The signal sequence of protein A functions efficiently as a secretory signal of human PTH in E.coli, but the hormone is preferentially cleaved at an internal site.

3. A small deletion in the signal sequence of protein A adjacent to the N-terminal part of PTH invalidates any subsequent processing, but is able to support a high efficiency secretory system. The uncleaved signal sequence stabilizes the hormone probably by imposing changes in the folding of the peptide.

4. In S.cerevisiae the mating factor α pheromone system functions in directing secretion of hPTH and gives rise to correct N-terminal processing of the hormone as well as internal cleavage.

5. Deletion of the spacer sequence, Glu-Ala-Glu-Ala, which is the processing site for the STE 13 enzyme, gives a more efficient N-terminal processing of human parathyroid hormone, but does not protect against internal degradation.

6. In E.coli and in S.cerevisiae, internal degradation of parathyroid hormone occurs at the same site after lysine in position 26.

7. Both in E.coli as well as in S.cerevisiae, other potential cleavage sites are much less susceptible to proteolysis, probably due to the inherent folding of the molecule.

ACKNOWLEDGMENT

This project has received economical support from the company Selmer-Sande A/S, from The Royal Norwegian Council for Scientific and Industrial Research (NTNF) (Grant PT 15.18471, and PT 15.15449) and from The Norwegian Research Council for Science and the Humanities (NAVF) P.no. 316.90/040. The project was also in part funded by a grant from the Nordic Yeast Research Program to OSG and from Anders Jahres Foundation for Promotion of Science.

REFERENCES

1. Cohn DV, Elting J (1983) In: Greep RO (ed) Recent Progess in Hormone Research, Academic Press, NY, pp. 181-209

2. Keutmann HT, Sauer MM, Hendy GG, O'Riordan JLH, Potts JT Jr (1978) Biochemistry 17:5723-5729

98

3.Hendy GN, Kronenberg HM, Potts JT Jr, Rich A (1981)   Proc Natl
Acad Sci USA 78:7365-7369

4.Cohn DV, MacGregor RR (1981)   Endocrine Rev 2:1-26

5.Potts JT Jr, Kronenberg HM, Rosenblatt M (1982)   Adv Protein Chem
35:323-396

6.Rosenblatt M (1984) In: Heam MTW (ed) Parathyroid hormone: intracellular transport, secretion, and receptor interaction, Marcel
Dekker, New York, pp. 209-296.

7.Gautvik KM, Gordeladze JO, Moxheim E, Gautvik VT (1984)   Eur Surg
Res 16:41-54

8.Høgset A, Blingsmo OR, Gautvik VT, Sæther O, Jacobsen PB,
Gordeladze OJ, Alestrøm P, Gautvik KM (1990) Biochem Biophys Res
Commun 166:50-60

9.Gabrielsen OS, Reppe S, Sletten K, Øyen TB, Sæther O, Høgset A,
Blingsmo OR, Gautvik VT, Gordeladze JO, Alestrøm P, Gautvik KM
(1990) Gene 90:255-262

10. Høgset A, Blingsmo OR, Saether O, Gautvik VT, Holmgren E,
Hartmanis M, Josephson S, Gabrielsen OS, Gordeladze JO, Alestrøm P,
Gautvik KM (1990) J Biol Chem 265:7338-7344

11. Graham FC, Van der Eb AJ (1973) Virology, 52:456-467

12. Maniatis T, Fritsch EF, Sambrook J (1982)   Molecular Cloning, A
Laboratory Manual, Cold Spring Harbor Laboratory Press, NY

13. Helfman DM, Hughes SH (1987)   Methods in Enzymology, 152:451-
458

14. Brake AJ, Merryweather JP, Coit DG, Heberlein UA, Masiarz FR,
Mullenbach GT, Urdea MS, Valenzuela P, Barr PJ (1984) Proc Natl
Acad Sci USA 81:4642-4646

15. Vlasu, GP, Bencen GH, Scarborough RM, Tsai P-K, Whang JL, Maack
T, Camargo MJF, Kirsher SW, Abraham JA (1986)   J Biol Chem 261:-
4789-4796

16. Zsebo KM, Lu H-S, Fieschko, JC, Goldstein L, Davis J, Buker K,
Suggs SV, Lai P-H, Bitter G (1986)   J Biol Chem 261:5858-5865

17. Loison G, Findeli A, Bernard S, Nguyen-Juilleret M, Marquet M,
Riehl-Bellon N, Carvallo D, Guerra-Santos L, Brown SW, Courtney M,
Roitsch C, Lemoine Y (1988) Bio/Technology 6:72-77

18. Bussey H (1988) Yeast 4:17-26

19. Fuller RS, Sterne RE, Thorner J (1988) Ann Rev Physiol 50:345-
362

20. Fisher JM and Scheller RH (1988) J Biol Chem 263:16515-   16518

21. Thomas G, Thorne BA, Thomas L, Allen RG, Hruby DE, Fuller R,
Thorner J (1988) Science 241:226-230

© 1991 Elsevier Science Publishers B.V. (Biomedical Division)
Site-Directed Mutagenesis and Protein Engineering
M.R. El-Gewely, editor.

SITE-DIRECTED MUTAGENESIS OF GLUTAREDOXIN FROM BACTERIOPHAGE T4

THORLEIF JOELSON, MATTI NIKKOLA, MARGARETA INGELMAN, OLOF BJÖRNBERG,

BRITT-MARIE SJÖBERG & HANS EKLUND

Department of Molecular Biology, Swedish University of Agricultural Sciences, Biomedical Center, Box 590, S-751 24 Uppsala, Sweden

INTRODUCTION

Thioredoxins and glutaredoxins are small redox-active proteins which have been isolated and characterized from widely different species. These proteins have one active site dithiol per molecule and their main function in cells is probably to reduce ribonucleotide reductase in the reduction of ribonucleotides to deoxyribonucleotides. Other functions have also been found[1].

Thioredoxins are distinguished from glutaredoxins by the way they are reduced. Thioredoxins are reduced by the enzyme thioredoxin reductase, an FAD containing enzyme that is reduced by NADPH. Glutaredoxins are reduced by glutathione. When *E. coli* is infected by bacteriophage T4, phage encoded glutaredoxin (earlier called T4 thioredoxin) and ribonucleotide reductase are produced. Functionally T4 glutaredoxin is a hybrid of *E. coli* thioredoxin and glutaredoxin and is reduced by both glutathione and thioredoxin reductase.

The amino acid sequences of several thioredoxins have been determined. The identity between molecules is high in the active site regions. In other parts of the molecules the homology is often low but a characteristic pattern of conserved residues exists. Most thioredoxins have no detectable amino acid homology to glutaredoxins which have a completely different pattern of conserved residues. The cysteine residues involved in the redox reactions are in all thioredoxins and glutaredoxins separated by two residues. Generally, thioredoxins have the sequence Cys-Gly-Pro-Cys in their active sites. The glutaredoxin sequences generally have the active site amino acid sequence Cys-Pro-Tyr-Cys. T4 glutaredoxin differs from the other glutaredoxins and has the active site sequence Cys-Val-Tyr-Cys.

The three-dimensional structures of T4 glutaredoxin and *E. coli* thioredoxin have been determined[2,3]. The two proteins have a similar fold despite the lack of sequence homology. The active site is located at the amino terminal end of a helix in both cases. The location of the active site in T4 glutaredoxin is shown in Figure 1.

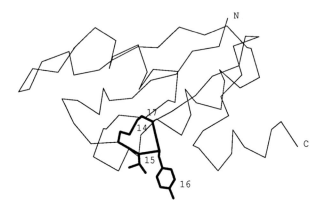

Fig. 1. Stereo drawing of the α-carbon chain of T4 thioredoxin. The amino acid residues at the active site are shown in thick lines.

We have initiated an investigation on structure and function of T4 glutaredoxin using site-directed mutagenesis. Since the residues between the cysteine residues are highly conserved within the thioredoxin and glutaredoxin families respectively, these residues were chosen as the primary targets for our studies[4]. In this study we have changed the active site residues of T4 glutaredoxin to mimic the local structure in this region of the thioredoxins and of the other glutaredoxins.

CLONING AND SITE DIRECTED MUTAGENESIS

DNA fragments from phage T4alc7 were cloned in M13mp9. The resulting library was screened with a radioactive probe constructed based on the amino acid sequence of T4 glutaredoxin. DNA from a phage that gave positive signal was cleaved and the resulting 1.8 kb fragment was cloned into the tetracycline resistance gene of pBR322. A 438 bp fragment containing the gene was cut out and cloned in M13mp18. In the next step the fragment was moved to the vector M13K19 for site-directed mutagenesis. The mutagenized fragments were cloned in the plasmid pMG524, which contains a heat inducible λP_L promoter for efficient expression of the mutant proteins. 25 - 50 mg protein were usually obtained from 12 g wet-weight bacteria.

Mutagenic primers contained 2-5 mismatches, and there were generally 9 bases flanking the mismatches at each side. The nucleotide sequence of the mutants were finally confirmed with plasmid sequencing.

ASSAYS

The disulfide reductase activity was determined using DTNB (5,5′-dithiobis-(2-nitrobenzoic acid) with the mutant proteins as substrates in the presence of thioredoxin reductase. Redox equilibria were determined essentially according to ref 4 and 5. The glutathione assays were performed as described by Holmgren[6,7].

ACTIVITY WITH THIOREDOXIN REDUCTASE

Our mutations of the active site residues (V15G, V15P, Y16P and V15G;Y16P) have rather small effects on the interactions with thioredoxin reductase. The activities of the mutant T4 glutaredoxins with thioredoxin reductase are similar to that of the wild type glutaredoxin. This indicates that a change to smaller residues in the investigated positions has little effect on enzyme - glutaredoxin interactions.

ACTIVITY WITH T4 RIBONUCLEOTIDE REDUCTASE

The rates of reaction for the mutant and wild-type glutaredoxins were determined with T4 ribonucleotide reductase. The rates were very similar for all mutants. The *E. coli* thioredoxin showed weak activity with T4 ribonucleotide reductase at the concentration of enzyme used in the T4 glutaredoxin assay system. An activity comparable to that of T4 glutaredoxin was reached at about an order of magnitude higher concentration of *E. coli* thioredoxin.

ACTIVITY WITH *E. COLI* RIBONUCLEOTIDE REDUCTASE

For the reaction with *E. coli* ribonucleotide reductase, which is the most specific of the enzymatic reactions, two of the mutants, V15G;Y16P and Y16P, gave 3-4 times lower apparent K_M, 30µM and 20µM respectively, than wild-type T4 glutaredoxin, 83µM. The catalytic efficiency (k_{cat}/K_M) of these two mutant proteins is significantly higher than that of the wild-type protein. The mutants were made more similar to *E. coli* thioredoxin in this respect and thus had one of the desired effects. The lowered K_M for the mutants may be interpreted as stronger binding to *E.coli* ribonucleotide reductase caused by the change of tyrosine to the less bulky hydrophobic side chain of proline.V15P and V15G had higher apparent K_M values than the wild-type protein.

REDOX POTENTIALS

A single change from Tyr to Pro in position 16 shifts the redox potential from -0.23 V to -0.25 V which is remarkably close to -0.26 V of the *E. coli* thioredoxin. Thus the redox potential for the Y16P mutant approaches the redox potential for the *E. coli* thioredoxin. The double mutant and the V15P mutant have intermediate redox potentials.

DOCKING OF GLUTATHIONE TO T4 GLUTAREDOXIN

Glutathione has been shown to reduce T4 glutaredoxin. We have modelbuilt glutathione bound to the crystallographically determined oxidized form of T4 glutaredoxin. There is a cleft in the vicinity of the active site disulfide in T4 glutaredoxin which is suitable for glutathione binding. *E. coli* thioredoxin does not have a similar cleft at its active site, only a shallow depression. This may explain why *E. coli* thioredoxin is not reduced by glutathione at concentrations suitable for glutaredoxins.

In oxidized T4 glutaredoxin only one cysteine, Cys-14 of the disulfide bridge is accessible, while Cys-17 is buried. Thus, glutathione has only one possible target in the primary reduction step. In order for the glutathione sulfur atom to contact the Cys-14 sulfur atom of the disulfide, the glutathione molecule has to enter into the cleft along the surface. One side of this cleft is formed by residues 12, 14 and 16, and the other by residues 64-66. The glutathione molecule fits very well to this cleft. The charged glutathione molecule was oriented in the cleft according to the charge distribution around the cleft (Figure 2). Asp-80 is at one end of the cleft while His-12 is located at the opposite end. These charges will stabilize the correct orientation of the glutathione molecule by interactions of the α-amino group of its Glu residue with Asp-80 and the carboxyl end of the glutathione peptide with His-12.

Other features of this binding mode is that the side-chain of Tyr-16 forms extensive van der Waals interactions with glutathione and a possible hydrogen bond to the α-amino group of glutathione. The central cysteine residue of glutathione forms hydrogen bonds with the main chain of residue 65 in an anti-parallel manner. Both these hydrogen bonds can not be formed with a normal *trans*-peptide bond between residues 65 and Pro-66. The *cis*-peptide bond is thus advantageous not only for stability and folding of the protein (as has been shown for *E. coli* thioredoxin), but in T4 glutaredoxin also for forming favorable interactions between protein and glutathione. The sulfur atom of glutathione is by these interactions positioned such that it can attack the Cys-14 sulfur atom of the glutaredoxin disulfide bridge. It is likely that similar interactions also occur between T4 glutaredoxin and thioredoxin reductase.

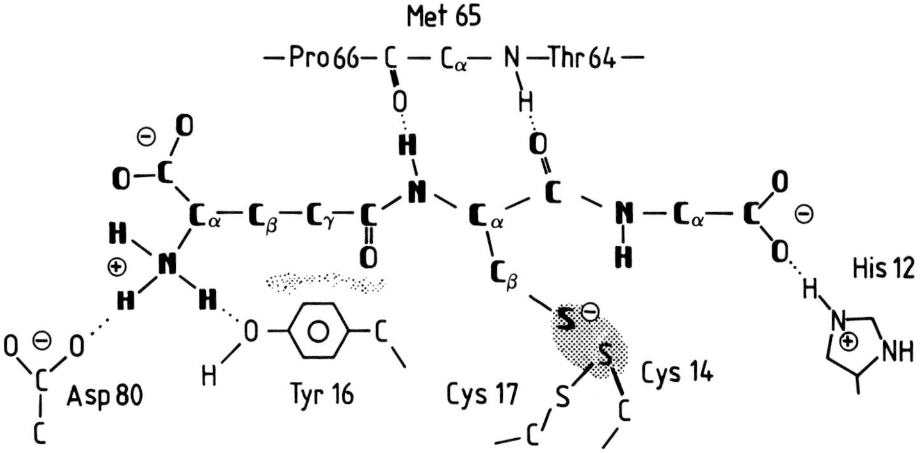

Fig. 2. Schematic drawing of the proposed glutathione - T4 glutaredoxin interactions. The hydrogen bonds are shown with dotted lines. We want to thank Ulla Uhlin for drawing the picture.

ACTIVITY WITH GLUTATHIONE

The set of site-directed mutant proteins which we constructed were all changed around the cleft described above. The ability of T4 glutaredoxin to function as a glutathione - disulfide oxidoreductase, i.e. as a transhydrogenase enzyme, was measured. The most marked deviation in reaction rate is observed for the mutant proteins which lack a Tyr in position 16. The wild-type T4 glutaredoxin functions 4 to 5 times better than these. The H12S substitution also gives a protein which work less well than the wild-type protein. In contrast, mutations at position 15 have no significant effect on the transhydrogenase activity of T4 glutaredoxin. *E. coli* thioredoxin shows no appreciable transhydrogenase activity.

In wild-type T4 glutaredoxin, the aromatic ring of Tyr-16 is proposed to contribute van der Waals contacts with the glutathione molecule and the hydroxyl group may further contribute to binding by forming a hydrogen bond to the α-amino group of the glutathione molecule. Glutaredoxin usually have Tyr at the corresponding position. Thioredoxins (e.g. the E.coli thioredoxin), which do not react with glutathione, have Pro in the corresponding position. The two mutant T4 glutaredoxins, which have a proline instead of Tyr-16, show reduced activity with glutathione and are thus more similar to E. coli thioredoxin in this respect.

The side chain of Val-15 is not directed towards the bound glutathione molecule but points away from it. The two mutant proteins with Pro or Gly substituting Val-15 behave as the wild-type T4 glutaredoxin in the glutathione reactions. It is thus consistent with the orientation of the Val-15 side-chain that a Gly or Pro substitution in this position should not change the interactions between the mutant proteins and the glutathione molecule.

The His-12 side-chain can interact electrostatically and via hydrogen bond with the carboxyl end of the extended glutathione molecule. In the H12S mutant protein, the hydroxyl group of the smaller serine side chain probably does not reach far enough to give a hydrogen bond. The partially charged interaction is also not possible in this mutant protein. Experimentally we find that the mutation H12S has smaller effects on the transhydrogenase activity than the Y16P mutations which implies that the interactions of His-12 contributes less than Tyr-16 to the interaction of T4 glutaredoxin with glutathione. The small effect of the substitution of His-12 to Ser in T4 glutaredoxin indicates that His-12 has no catalytic function in this protein.

CRYSTALLOGRAPHIC INVESTIGATION OF A MUTANT T4 GLUTAREDOXIN

The three-dimensional structure of one of the T4 glutaredoxin mutant proteins has so far been determined. The double mutant with the active site sequence Cys-Gly-Pro-Cys crystallizes in a new crystal form with one molecule in the asymmetric unit and diffract to at least 1.4 Å resolution. These crystals have allowed a structure determination of higher accuracy than the structure determination of the native protein. The structure of the mutant protein was solved with molecular replacement methods and subsequently refined using the X-plor program.

The overall structure of the mutant is very similar to the native protein. The largest differences occur in regions involved in different packing interactions. The torsion angles of the active site disulfide bridge are also very similar in the mutant and wild-type proteins.

DISCUSSION

The present study investigates the extent to which the residues at the active site influence the activity of thioredoxin in its redox-reactions. We have modified the active site of T4 glutaredoxin to mimic the local structure in other thioredoxins and glutaredoxins. Such changes will of course not give perfect resemblance to the other proteins since the surroundings of the disulfide bridge differ. The interaction areas for thioredoxin reductase and ribonucleotide reductase most probably extend beyond the four active site residues. Other residues must contribute to binding or catalysis. In *E. coli* thioredoxin, the residues of the disulfide bridge form a short helix, which is tilted relative to the following long helix due to a kink caused by Pro-40. In T4 glutaredoxin this is a long straight helix. Furthermore, loops around the active site differ in length between T4 glutaredoxin and *E. coli* thioredoxin. Mutations of *E. coli* thioredoxin demonstrate that residues far away in the primary structure but close to the active site area strongly affect its activity[8]. Residues in this area around the active site differ between members of the thioredoxin family and even more so for glutaredoxins.

Structural comparisons of T4 and *E. coli* thioredoxins have led to the identification of a hydrophobic surface on one side of the disulfide bridge which was suggested to interact with thioredoxin reductase and ribonucleotide reductase. The main structural difference between T4 glutaredoxin and *E. coli* thioredoxin in this region is Tyr-16. The side chain of Tyr-16 protrudes into solvent in T4 glutaredoxin whereas the surface of the *E. coli* thioredoxin is flat.

There is a preference for Pro in the sequences of the active site tetra-peptide of thioredoxins and glutaredoxins, and also a preference for Gly in the thioredoxins. Obvious questions are how these residues affect the stability of the active site, and the relationship between their structure and activity. Residues 15 and 16 in T4 glutaredoxin adopt helical conformation. Pro in either of these positions need not change the conformation of the main-chain, prolines are common in the first turn of a helix. Although Pro can easily fit in positions 15 and 16 in structures similar to that of oxidized wild-type thioredoxin, it would make the main-chain more rigid. As a result, the switch from

106

one state to the other would be hindered. In contrast, a Gly residue, which in most thioredoxins is present in the equivalent position to 15, would impart greater flexibility to the main-chain.

In the reduced form of thioredoxin, the active site has to adopt a more open conformation since the disulfide bridge is broken. A proline in position 16 would make the main-chain more rigid than any other residue type in this position. It should stabilize the helical conformation of the active site peptide and thus favor the oxidized form of the protein. Consistent with our hypothesis we observe a lower redox equilibrium for mutant T4 glutaredoxins with Pro at position 16. This effect can be partly reduced with the more flexible Gly in the preceding position. Position 15 is the first residue of the helix and has larger flexibility than position 16. A mutant with Pro in this position would not be expected to affect the redox potential to the same extent, which is consistent with our results.

CONCLUSIONS

In conclusion, our results show that the redox properties of T4 glutaredoxin and its interactions with *E. coli* ribonucleotide reductase can be made to mimic those of the *E. coli* thioredoxin by mutating the residues between the two active site cysteines to the sequence found in the bacterial protein whereas these changes do not significantly affect its interaction with T4 ribonucleotide reductase and *E. coli* thioredoxin reductase. Generally a change in the second position investigated has a larger effect on activity than the changes in the first position. The mutant T4 glutaredoxins, where Tyr-16 has been substituted have kinetic parameters which deviate from those of the wild-type T4 glutaredoxin. Glutaredoxins which were changed in position 15 behave more or less like the wild-type protein. The exception is in the *E. coli* ribonucleotide reductase reaction, where both mutants give higher apparent K_M compared to the wild-type T4 glutaredoxin.

ACKNOWLEDGEMENTS

This work was supported by a special grant from the Swedish Ministry of Agriculture, by the Swedish Natural Science Research Council and by the Swedish National Board for Technical Development.

REFERENCES

1. Holmgren, A. (1985) Ann. Rev. Biochem. 54: 237-271

2. Söderberg, B.-O., Sjöberg, B.-M., Sonnerstam, U. & Brändén, C.-I. (1978) Proc. Natl.
    Acad. Sci. USA 75: 5827-5830

3. Holmgren, A., Söderberg, B.-O., Eklund, H. & Brändén, C.-I. (1975) Proc. Natl. Acad.
    Sci. USA 72: 2305-2309

4. Joelson, T., Sjöberg, B.-M. & Eklund, H. (1990) J. Biol. Chem.

5. Berglund, O. & Sjöberg, B.-M. (1970) J. Biol. Chem. 245: 6030-6035

6. Holmgren, A. (1978) J. Biol. Chem. 253: 7224-7430

7. Holmgren, A. (1985) Methods in Enzymology 113: 525-540

8. Model, P. & Russel, M. (1986) in Thioredoxin and Glutaredoxin Systems, Structure
    and Function (Holmgren, A., Brändén, C.-I., Jörnvall, H. & Sjöberg, B.-M. eds.), Raven
    Press, New York, pp. 323-329

© 1991 Elsevier Science Publishers B.V. (Biomedical Division)
Site-Directed Mutagenesis and Protein Engineering
M.R. El-Gewely, editor.

MUTATIONAL ANALYSIS OF A TRANSCRIPTIONAL REGULATORY PROTEIN OF BACTERIOPHAGE P2

KIRSTI GEBHARDT AND BJORN H. LINDQVIST

Institute of Biology and The Biotechnology Centre of Oslo, University of Oslo, P.O. Box 1125 Blindern, 0316 Oslo 3, (Norway)

INTRODUCTION

Bacteriophage P2 and its relatives provide a convenient and well defined system for the study of gene regulation in general and viral transactivation in particular (1). Synthesis of the late morphogenetic head and tail components of bacteriophage P2 is positively controlled by the Ogr protein (2). It is coded for by the P2 *ogr* gene (3,4) which is expressed between the early and the late transcription during a P2 infection of *E. coli* (Birkeland et al., manuscript in preparation). Ogr acts by facilitating *E. coli* RNA polymerase to initiate transcription at a set of four late promoters (Fig. 1). These differ substantially from the *E. coli* consensus promoter and they are inactive in the absence of Ogr function (2,5,6). In addition, Ogr dependent transcription requires P2 DNA replication (6).

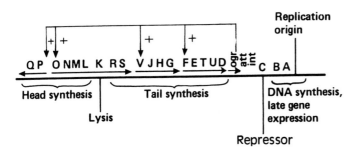

Fig. 1. Genetic map of bacteriophage P2. The 4 late operons positively regulated by Ogr and their direction of transcription are marked with arrows, as is the *ogr* gene.

The Ogr protein consists of 72 amino acids (MW 8.300) and is believed to interact with DNA in the vicinity of the promoters in order to position the host RNA polymerase for transcription initiation (3,4). In fact the amino acid sequence of the Ogr protein predicts two potential Zn-finger structures (8), one

Cys$_2$-Cys$_2$ and one Cys$_2$-His$_2$ type (9,10), which could be involved in DNA binding. We have ruled out the Cys$_2$-His$_2$ type as a functional Zn-finger by site-directed mutagenesis (Gebhardt and Lindqvist, manuscript in preparation). Furthermore, studies with purified Ogr protein have shown that Zn^{2+} binds to the protein, most likely through Cys$_2$-Cys$_2$ coordination (11). Attempts to demonstrate direct binding of Ogr to P2 DNA have so far failed, however (12,13).

The mutation that initially defined the *ogr* gene was discovered when a high titer stock of P2 was plated on an *E. coli* mutant, called RpoA109 (14). The *rpoA109* mutation leads to a Leu to His substitution in the α subunit of the host RNA polymerase (15). This mutation effectively blocks P2 late transcription. P2 mutants able to overcome this block were isolated at an extremely low frequency (10^{-10}) (14). Based on the compensatory nature of the mutation in *ogr*, the α subunit(s) of RNA polymerase and the Ogr protein are thought to interact during transcriptional activation of the P2 late promoters. Additional mutations in the α subunit which affect transcription of other positively regulated *E. coli* genes have been described (16,17,18,19). Two of these mutations do not affect growth of phage P2 (16,18), while the others have not been tested. Hence, Ogr probably represents just one example of a more general class of α subunit specific gene transactivators in *E. coli*. In this communication we investigate aspects of the assumed Ogr-RNA polymerase interaction by a mutational analysis the Ogr protein.

MATERIALS AND METHODS

A. Cloning and sequencing of P2 Ogr mutants:

Phage strains: P2 *vir1ogr8* (14), P2 *vir1amE30ogr9* (Bertani, G. unpubl.), P2 *vir1amA127ogr11* and P2 *del6c5ogr23* (Haggård-Ljungquist, unpubl.). *E. coli* C strains C-2124 (*rpoA109*) (14) and C-1792 (*supF*) (20) were used as hosts for P2*ogr8* and *ogr23*, and for P2*ogr11*, respectively. P2*ogr9* was obtained as DNA. Culture media were L broth and L agar (21). Phage stocks and phage DNA were prepared essentially as described in (21) and (22). Standard techniques for recombinant DNA methodology were used (23). Pure phage DNA was digested with NruI and PstI and cloned into M13mp18 (SmaI/PstI) (24). Protocols for M13 growth and DNA isolation are as described in (25). Clones with the correct *ogr*-containing fragment (1.8kb) were isolated by size determination in agarose gels and sequenced according to Sanger (26).

B. Site-directed *in vitro* mutagenesis:

The uracil-incorporating method of Kunkel et al. (27) was used. Bacterial strains, reagents and protocols were supplied by the Phagemid *in vitro* Mutagenesis Kit (Bio-Rad Laboratories). A 590 bp *ogr* wt fragment, kindly provided by N.K. Birkeland (from P2*vir22*, ClaI/AatII, via pSP64, EcoRI/HindIII) was cloned into the Bluescript SK+ vector (Stratagene). This construct is referred to as pKG100. Single stranded template DNA was isolated by infection with a helper phage; R408 (28) or M13K07 (29). Mutagenic oligonucleotide primers were synthezised in our laboratory using a Pharmacia or Beckman DNA syntheziser. The mutations were verified by sequencing.

C. Complementation tests/one step burst size determinations:

pKG100 containing the different *ogr* mutations were transformed into *E. coli* C-1a (30). The transformants were infected with the *ogr*-defective phage P2*del15* (31) using a multiplicity of infection of 0.01. *E. coli* C-1055 (32) containing the *ogr*-expressing construct pKG100 was used as indicator strain. Procedures for one-step burst size determination were essentially as described in (31). *E. coli* C strains harboring Bluescript plasmids show heat sensitivity and were therefore grown at 30°C.

RESULTS

The *ogr* mutations are identical:

The fact that the compensatory Ogr mutants appear at such a low frequency suggests that the mutation leads to a very specific change at the amino acid level. Two independent *ogr* mutations have previously been characterized (*ogr1* and *52*), and in both cases a Tyr residue at position 42 is replaced by a Cys (TAT to TGT) (3,4). We have sequenced four additional independent *ogr* mutations (*ogr 8, 9, 11* and *23*) and all turn out to have the very same change as the two previously characterized (Fig. 2). The fact that only Cys appears in the Ogr mutants suggests that the compensatory effect with regard to amino acid replacement is limited. One-base changes in codon 42 could, in addition to Cys, give rise to His, Asn, Asp, Phe and Ser. The absence of these amino acids among the Ogr mutants seems to rule out these as compensatory amino acids in the RpoA109 interplay. We cannot, however, rule out other amino acids at position 42 or elsewhere in the protein which require more than a single nucleotide change.

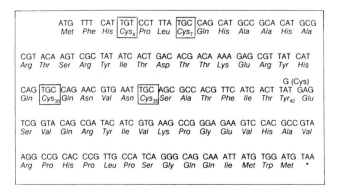

Fig. 2.   The *ogr* gene and its amino acid sequence.   The six *ogr* mutations characterized so far (*ogr 1, 8, 9, 11, 23* and *52*), are all identical Tyr to Cys substitutions at position 42.   Cys residues believed to coordinate Zn^{2+} are boxed (11).

Mutational analysis of the $CysX_3Phe$ element:

We have observed that a small sequence element, CX_3F, which is conserved between Ogr and its two functional equivalents (Delta and B) from the related bacteriophages P4 and 186 (33,34), also appears as a conserved element in other Cys-rich proteins and protein families (Table 1).   In these cases CX_3F is always preceded by another Cys (2-4 residues away) and a hydrophobic, usually aromatic, residue (4-7 residues away).   This extended sequence or motif is in fact identical to the first half of the TFIIIA-like Zn-finger consensus sequence (9).   As evident from table 1, the proteins show diverse functions but in all cases the motif seems to be involved in molecular recognition, either DNA-protein or protein-protein interactions.   In the case of the first four protein groups in table 1 the Cys residues of the motif are also known (or assumed) to serve as two of four Zn-ligands in a tetrahedrally arranged Zn-complex (11,35,36,37).

The 3-dimensional structure has been solved for a TFIIIA-like Zn-finger peptide and for *E. coli* aspartate transcarbamoylase (ATCase) (38,36).   In both cases the motif Y/F/L X C X_{2-4} C X_3 F has been shown to adopt an anti-parallel ß-sheet configuration (ß-hairpin).   The flanking hydrophobic residues are packed against other hydrophobic amino acids of the protein/peptide and they also have a role in stabilizing the ß-sheet through reciprocal (backbone) hydrogen bonding. In fact, Berg (39) was able to successfully predict the structure of a TFIIIA-like

Zn-finger based on the known 3-D structure of ATCase and by recognizing a similar arrangment of cysteins and hydrophobic residues in the metal binding site of the two proteins. The Zn-finger of TFIIIA is involved in DNA-protein interaction, while in ATCase the Zn-binding area mediates protein-protein contacts between the regulatory and catalytic subunits of the enzyme (36). CX_3F is also present in all isolates of the Tat protein of HIV-virus (8,40) and in the EGF-family consensus (41). However, for these proteins the pattern is part of a larger Cys-rich domain with its own typical consensus.

Protein type:	Protein function	Motif:	Motif function	Zn-binding	Secondary structure
Ogr, Delta and B (P2-like phages)	Transcription factors	C X_2 C X_{19} Y/F X_2 C X_4 C X_3 F	Zn-binding, DNA-binding and/or RNA pol. binding	+	?
TFIIIA Zn-fingers (C_2-H_2 type)	Transcription factors	Y/F X C X_{24} C X_3 F X_5 L X_2 H X_3 H	Zn-finger DNA-binding	+	antiparallell beta sheet
Aspartate transcarbamoylase	Biosynthetic enzyme	C X_4 C X_{21} L X C X_2 C X_3 F	Zn-binding, contact between R/C subunits	+	antiparallell beta sheet
1.finger, Steroid hormone receptor family (C_2-C_2 type)	Transcription factors	C X_2 C X_{11} hp X C X_2 C X_3 F	Zn-finger DNA-binding	+	?
CONSENSUS:		Y/F X C X_{2-4} C X_3 F			
Tat (HIV)	Transcription factor	Y/F X_4 C X_2 C X_3 F	Zn-binding Dimerization	+	?
EGF-family	Variable	Y/F X C X C X_3 F/Y	Variable	?	?

Table 1. Alignment of proteins matching the Y/F X C X_{2-4} C X_3 F consensus. Known or putative Zn-ligandig amino acids are underlined. Abbreviations: hp=hydrophobic residue, C=Cys, F=Phe, H=His, L=Leu, Y=Tyr. X=position of an amino acid (may be conserved, but the identity is omitted for clarity). References: Aspartate transcarbamoyl transferases (regulatory subunits): E. coli: (42), S. typhimurium: (43). Steroid hormone receptors: (37,44). EGF family (proteins containing epidermal growth factor-like repeats): (41). HIV Tat proteins: (40,45).

The CX_3F sequence appears to be the most highly conserved component of the consensus sequence described above. Its presence in all these proteins, some of which are transcriptional activators prompted us to investigate its role in the Ogr protein. By site directed mutagenesis we have changed the Cys_{35} to Gly and introduced a set of 7 point mutations in the codon of Phe_{39} of the *ogr* gene. For this purpose the *ogr* gene was cloned into the Bluescript vector, now designated as pKG100. As can be seen in table 2, only Leu and Ile at position Phe_{39} retain some Ogr function. All other changes, including Cys_{35} to Gly result

114

in total loss of function as measured in a complementation test. E.g. Phe to Tyr, which only represents the addition of a hydroxyl group to the ring structure of phenylalanine destroys Ogr function. The fact that only Leu and Ile are tolerated suggests that Phe$_{39}$ is involved in hydrophobic interactions. We have also exchanged one of the intervening residues in CX$_3$F, (Ala$_{37}$ to Cys) which is **not** conserved between Ogr and its functional equivalents, Delta and B. This mutation has minimal effect on Ogr function and highlights the conserved nature of the Cys and Phe residues in CX$_3$F.

Mutation	Burst size (% of wildtype)
Cys 35 to Gly	< 0.1
Ala 37 to Cys	82.7
Phe 39 to Ser	< 0.1
" to Asp	< 0.1
" to Tyr	< 0.2
" to Cys	< 0.1
" to Pro	< 0.1
" to Ile	12.0
" to Leu	33.4
Tyr 42 to Cys	46.0
Tyr 42 to Cys + Phe 39 to Leu	< 0.1
Glu 43 to Gly	< 0.1
Negative control	< 0.1

Table 2. *In vitro* made mutations in the Ogr protein and their effect on Ogr function. Mutations are listed in order of apparance from the N-terminus. Mutant Ogr proteins were expressed from pKG100 and tested for ability to complement the *ogr*-defective phage P2*del*15 (31) in one-step burst size experiments. Burst size values are given as % of wt Ogr function and represent an average of two or more independent determinations. Cells containing vector without the *ogr* gene served as negative control.

The CX$_3$F sequence in Ogr is positioned only three amino acids away from the previously described *ogr* mutation at position 42, which raises the possibility that CX$_3$F also takes part in the presumed interaction with one or both of the α subunits of RNA polymerase. We have introduced the *ogr* mutation (Tyr$_{42}$ to Cys) into the pKG100 construct, both separately and in combination with the Phe$_{39}$ to Leu mutation. A loss of Ogr function is observed when these otherwise functional mutations are combined together. This result could reflect a general distortion of the Ogr structure, but the differential effect of Phe$_{39}$ to Leu when present alone as opposed to when in combination with the *ogr* mutation, is indicative of CX$_3$F being part of the domain which participates in the RNA polymerase interaction.

Finally, we have exchanged the acidic Glu$_{43}$ (present in Ogr, Delta and B), with a Gly, primarily for its close proximity to the *ogr* mutation at position 42. As can be seen from table 2 this alteration leads to a complete loss of Ogr function.

DISCUSSION

Our analysis of the *ogr* gene and its product has focused on the amino acids in the 35-43 positions. The compensatory effect of the *ogr* mutation at position 42 (Tyr to Cys) on the *rpoA109* allele of the α subunit of *E. coli* RNA polymerase appears to be highly specific. All six *in vivo* isolated *ogr* mutations analysed so far are identical one-base substitutions at codon 42. This result suggests that the compensatory effect is unlikely to be achieved by other amino-acid substitutions involving one-base changes. At codon 42 this rules out His, Asp, Asn, Phe and Ser as compensatory amino acids. Recently, *in vitro* mutagenesis studies have shown that also Ala and probably Gly can compensate for Tyr at position 42 (Anders et al., manuscript in preparation), both of which require more than a single nucleotide change.

As shown in table 1 the CX$_3$F element is part of the larger consensus sequence Y/F X C X$_{2-4}$ C X$_3$ F which has been shown to constitute an antiparallel ß-sheet in the two functionally distinct proteins TFIIIA* and aspartate transcarbamoylase**. In the case of TFIIIA, the Zn-finger is formed by Zn-binding between two cysteins at the base of the ß-hairpin and two histidine residues, 9 and 12 amino acids further to the **right** (Table 1). In the case of ATCase, Zn-binding takes place through the cysteins of the ß-sheet and another cysteine pair, 27 and 32 residues to the **left** of the CX$_3$F element. Both types of Zn-binding motifs, in which the antiparallell ß-sheet module is central, represent domains which appear designed for either DNA-protein (TFIIIA-like) or protein-protein (ATCase) interactions. Although the ß-sheet module is common to both motifs, they differ in the configuration surrounding the **second** pair of Zn-ligands. In the case of TFIIIA the pair of histidines occur in an α-

---

* only demonstrated for a synthetic TFIIIA-like 30 residue peptide (38)
** in ATCase, the first residue of the motif is a Leu (hydrophobic)

helix configuration while for ATCase another ß-hairpin is used (Cys-Cys) (36,38,39).

It seems reasonable to suggest that the near perfect consensus sequence Y/F X_2 C X_4 C X_3 F of the Ogr, Delta and B proteins adopts a similar anti-parallel ß-sheet structure to that of ATCase and TFIIIA. This assumption is supported by the results of our mutational analysis of the CX_3F element, which is the most highly conserved component of this consensus sequence. In particular, the observation that only hydrophobic residues seem to be tolerated at the Phe_{39} position is consistent with the role depicted for this residue at the corresponding position in ATCase and TFIIIA.

Furthermore, the arrangment and spacing of assumed Zn-binding amino acids is almost identical in Ogr and ATCase (two cysteine pairs spaced by 22 vs. 23 amino acids). This similarity, and the fact that a TFIIIA-like Zn-finger has been ruled out by mutational analysis, implies that the cysteins of the proposed ß-sheet in Ogr interacts (through Zn-binding) with the cysteine residues located 28 and 31 residues to the left of the CX_3F sequence, like in ATCase (see Table 1). Such a structural domain is likely to be essential for Ogr, Delta and B function and explains the conserved nature of the amino acids involved for the three related gene regulators. The similarity of the Zn-binding domain of ATCase and that of Ogr had been noted earlier by Gebhardt (8) and has also been pointed out by Lee (13). If the Zn-binding domain of Ogr, like that of the ATCase, is used exclusively in protein-protein interaction, the CX_3F sequence would probably take part in the assumed interaction with RNA polymerase, and in particular with the α subunit.

ACKNOWLEDGMENTS:

We would like to thank Gail Christie for communicating unpublished results and for commenting on the manuscript. We are also grateful to Mel Sunshine, Ole Jørgen Marvik and Timothy Lavelle for helpful comments. This work was supported by the Norwegian Research Council for Science and Humanities, grants 451.89/038 and 451.90/012.

REFERENCES

1. Bertani, L.E. and Six, E.W. (1988) In: The Bacteriophages, Vol.2., ed. R. Calendar 73-143 New York: Plenum Publishing Corporation.

2. Grambow, N.G., Birkeland, N.K., Anders, D.L., Christie, G.E. (1990) Gene, in press

3. Birkeland, N.K. and Lindqvist, B.H. (1986) J. Mol. Biol. 188, 487-490

4. Christie, G.E., Haggård-Ljungquist, E., Feiwell, R., Calendar, R. (1986) Proc.Natl. Acad. Sci. USA 83, 3238-3242

5. Christie, G.E. and Calendar, R., (1983) J. Mol. Biol. 167, 773-790

6. Christie, G.E. and Calendar, R., (1985) J. Mol. Biol. 181, 373-382

7. Geisselsoder, J., Mandel, M., Calendar, R., Chattoray, D.K. (1973) J. Mol. Biol. 77, 405-415

8. Gebhardt, K., (1988) Cand. scient thesis, University of Oslo, Norway

9. Klug, A. and Rhodes, D. (1987) Trends. Biochem. Sci. 12, 464-469

10. Evans, R.M. and Hollenberg, S.M. (1988) Cell 52, 1-3

11. Lee, T. -C. and Christie, G.E. (1990) J. Biol. Chem. 265, 7472-7477

12. Kristiansen, E. (1987) Cand.scient thesis, University of Oslo, Norway

13. Lee, T. -C., (1989) Ph.D. Thesis, Virginia Commonwealth University, Richmond, Virginia, USA

14. Sunshine, M.G. and Sauer, B. (1975) Proc. Natl. Acad. Sci. USA 72, 2770-2774

15. Fujiki, H., Palm, P., Zillig, W., Calendar, R., Sunshine, M. G. (1976) Mol. Gen. Genet. 145, 19-22

16. Rowland, G.C., Giffard, P.M., Booth, I.R. (1985) J. Bacteriol. 164, 972-975

17. Giffard, P.M. and Booth, I.R. (1988) Mol. Gen. Genet. 214: 148-52

18. Garret, S., Silhavy, T.J., (1987) J. Bacteriol. 169: 1379-85

19. Matsuyama, S. and Mizushima, S. (1987) J. Mol. Biol. 195, 847-853

20. Sunshine, M.G., Thorn, M., Gibbs, W., Calendar, R., Kelly, B. (1971) Virology 46, 691-702

21. Bertani, L.E. and Bertani, G. (1970) J. Gen. Virol. 6, 201-212

22. Lindqvist, B.H. (1981) Gene 14, 231-241

23. Maniatis, T., Fritsch, E.F., Sambrook, J. (1989) Molecular Cloning. A laboratory manual. Second edition. Cold Spring Harbor Laboratory Press

24. Messing, J.V. (1983) Meth. in Enzymol. 101, 20-78

25. Carolina Workshops on Sequencing and Mutagenesis 1988, manual

26. Sanger, F., Nicklen, A.R., Coulson, A. (1977) Proc. Natl. Acad. Sci. USA 74, 5463-5677

27. Kunkel, T.A., Roberts, J.D., Zakour, R.A. (1987) Methods Enzymol. 154, 367-382

28. Russel, M., Kidd, S., Kelley, M.R. (1986) Gene 45, 333-338

29. Vieira, J. and Messing, J.V. (1987) Meth. Enzymol. 154, 3-

30. Sasaki, I. and Bertani, G. (1965) J. Gen. Microbiol. 40, 365-376

31. Birkeland, N.K., Christie, G.E. and Lindqvist, B.H., (1988) Gene 73, 327-335

32. Sironi, G. (1969) Virology 37, 163-176

33. Kalionis, B., Pritchard, M., Egan, J.B. (1986b) J. Mol. Biol. 191, 211-220

34. Halling, C., Sunshine, M.G., Lane, K.B., Six, E.W., Calendar, R. (1990) J. Bacteriol. 172, 3541-3548

35. Miller, J., McLachlan, A.D., Klug, A. (1985) EMBO J. 4, 6, 1609-1614

36. Monaco, H.L., Crawford, J.L., Lipscomb, W.N. (1978) Proc. Natl. Acad. Sci. USA 75, 11, 5276-5280

37. Freedman, L.P., Luisi, B.F., Korszun, R., Basavappa, R., Sigler, P.B., Yamamoto, K.R. (1988) Nature 334, 543-546

38. Lee, M.S., Gippert, G.P., Soman, K.V., Case, D.A., Wright, P.E. (1989) Science, 245, 635-637

39. Berg, J.M., (1988) Proc. Natl. Acad. Sci. USA 85, 99-102

40. Meyers, (1988) In: Human retroviruses and Aids. Los Alamos National Laboratory, Los Alamos, NM

41. Rothberg, J.M., Hartley, D.A., Walther, Z., Artavanis-Tsakonas, S. (1988) Cell 55, 1047-1059

42. Schachman, H.K., Pauza, C.D., Navre, M., Karels, J.J., Wu, L., Yang, Y.R. (1984) Proc. Natl. Acad. Sci. USA 81, 115-119

43. Michaels, G., Kelln, R.A., Nargang, F.E. (1987) Eur. J. Biochem. 166, 55-61

44. Evans, R.M. (1988) Science 240, 889-895

45. Frankel, A.D., Bredt, D.S., Pabo, C.O. (1988) Science 240, 70-73

© 1991 Elsevier Science Publishers B.V. (Biomedical Division)
Site-Directed Mutagenesis and Protein Engineering
M.R. El-Gewely, editor.

IN VITRO MUTAGENESIS OF RAT CATECHOL-O-METHYLTRANSFERASE

KENNETH LUNDSTRÖM, HANNELE AHTI AND ISMO ULMANEN
Orion Corporation, Laboratory of Molecular Genetics, Valimotie 7,
00380 Helsinki, Finland

INTRODUCTION

Catechol-o-methyltransferase (COMT) catalyses the transfer of a
methyl group from S-adenosyl-L-methionine to one of the phenolic
hydroxyl groups of a catechol substrate. This physiologically important function inactivates catechol hormones, catecholamine
neurotransmitters and many neuroactive drugs. To increase the efficiency of these drugs, like L-dopa in Parkinson's disease, specific
inhibitors against COMT have been developed. For molecular analysis
and to study the interaction between COMT inhibitors and the enzyme, we have cloned the rat liver COMT gene in our laboratory[1]. Sequencing revealed an open reading frame of 663 nucleotides encoding
a 221 amino acid long polypeptide. After introducing the coding region of the rat COMT gene into an E. coli expression vector, soluble and active recombinant COMT could be produced[2].

The aim of the present study has been to increase the knowledge
of the structure of the COMT enzyme. A series of 5′ end deletions
and site-specific mutations of the COMT gene have been made using
the PCR technique. Here we present the expression of the mutant
COMT enzymes in E. coli and preliminary characterization in SDS-
PAGE, Western blotting and a COMT activity assay.

MATERIALS AND METHODS

Expression of recombinant proteins. Protein production in E. coli
was carried out using vector pKEX14 (Fig. 1B). The vector utilizes
the T7 promoter. E. coli JM109 (DE3) cells (Promega) containing mutant constructs were grown and induced for 3h at +37°C with IPTG.
Cells were pelleted, lysed and subjected to gel electrophoresis or
COMT activity assay.

In vitro mutagenesis. 5′ end deletions of the COMT gene were obtained by using oligonucleotide primers with an EcoRI site, the ATG
initiation codon and sequences homologous to various parts of the
5′ end region. The 3′ end primer contained a Hind III site, the natural translation terminator TGA and sequences homologous to the

120

ultimate 3′ end coding region. PCR fragments were amplified under
conventional reaction conditions and inserted into the expression
vector pKEX14. Site-specific mutagenesis was carried out applying
the Megaprimer technique[3]. Amplified mutagenised fragments were
cloned into the bacterial expression vector. The accuracy of the
mutants was confirmed by DNA sequencing. The selection of amino
acid substitutions was based on structure-function studies using
COMT inhibitors[4] and data proposing that at least one sulfhydryl
group is present in the active site[5].

RESULTS
   DNA constructs for 5′ end deletions were obtained by applying the
PCR technique. Three different constructs with deletions of 10, 33
and 68 amino acids (Fig. 1A) were introduced into the E. coli expression vector.
   Site-specific mutants were constructed by amplifying the mutated
region and using the amplified fragment as a megaprimer for a second PCR synthesis to recieve the full-length mutated coding reg-

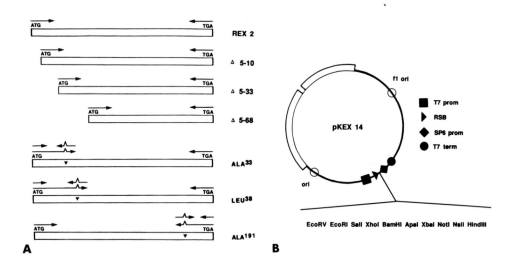

Fig. 1.A) Constructs made for mutagenesis studies. Arrowheads (→),
localization and direction of oligonucleotide primers used for PCR
reactions; bent arrowheads (⌐→), oligonucleotides and "megapri-
mers" containing site-specific mutations; triangles (▼) localization of mutations; ATG, initiator Met; TGA, translation terminator.
B) E. coli expression vector pKEX14.

ions (Fig. 1A). The first Cys^{33}, assumed to form a disulphide bridge with Cys^{191}, was substituted by an Ala^{33}, Trp^{38} was exchanged to Leu^{38} and the last Cys^{191} to Ala^{191}.

The size and amount of E. coli produced proteins were studied in SDS-PAGE (Fig. 2A). A roughly equal expression level of COMT protein could be seen for all constructs. Δ5-33 and Δ5-68 showed a slightly elevated expression. As expected, the deletion constructs expresssed proteins which had a faster mobility. The site-specific construct Ala^{33} showed protein of the same size as the native COMT, but Leu^{38} had surprisingly a slightly slower mobility.

The immunoreactivity of the mutated COMT proteins was studied in Western blots (Fig. 2B). It could be shown that all deletion constructs and site-specific mutants reacted with anti-COMT serum at comparable levels to the native protein.

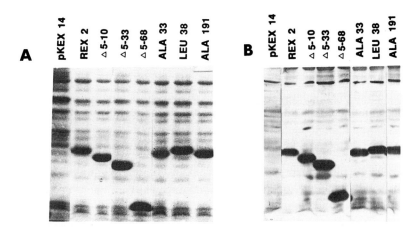

Fig. 2. SDS-PAGE (10%) analysis[6] of protein production in E. coli. A) Coomassie Brilliant Blue staining. B) Western blotting using polyclonal anti-COMT antiserum (1:200) and Protoblot Western Blot AP System (Promega).

Lysed bacteria showed no or extremely low levels of COMT activity for the deletion mutants (Fig. 3). Interestingly a 10 amino acid deletion was sufficient to almost completely abolish the activity. For the site-specific mutations, Ala^{33} and Leu^{38} substitutions led to the production of totally inactive protein, although the expression levels were equal to that for the wild type COMT. However, the Ala^{191} had only little or no effect on the COMT activity.

122

For further characterization of the COMT enzyme, 3' end deletions are under preparation, which will provide some information of the importance of the C-terminal region of the enzyme.

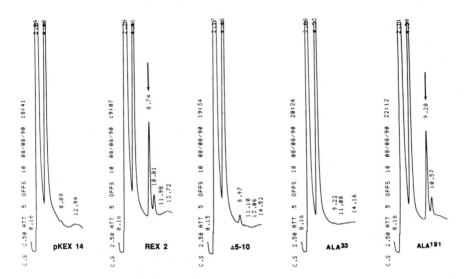

Fig. 3. COMT activity assay carried out according to the HPLC detection method by Nissinen and Männistö[7]. Arrowheads (→) stand for COMT peaks.

ACKNOWLEDGEMENTS

We are grateful to Dr. Johan Peränen (Institute of Biotechnology, Helsinki) for kindly providing the original expression vector and Dr. Jyrki Taskinen for critical reading of the manuscript. We wish to thank Marina Bengtström and Maritta Putkiranta for the expert synthesis of oligonucleotides and Marja Jänne, Raija Savolainen and Seija Ukkonen for excellent technical assistance.

REFERENCES
1. Salminen M, Lundström K, Tilgmann C, Savolainen R, Kalkkinen N, Ulmanen I (1990) Gene, in press
2. Lundström K, Peränen J, Jalanko A, Tilgmann C, Ulmanen I (1990) Manuscript in preparation
3. Sarkar G, Sommer SS (1990) Biotechniques 8 (4) 404-407
4. Taskinen J, Vidgren J (1990) Unpublished results
5. Borchardt RT, Thakker DR (1976) Biochim. Biophys. Acta 445, 598-609
6. Laemmli UK (1970) Nature 227, 680-685
7. Nissinen E, Männistö P (1984) Anal. Biochemistry 137, 69-73

© 1991 Elsevier Science Publishers B.V. (Biomedical Division)
Site-Directed Mutagenesis and Protein Engineering
M.R. El-Gewely, editor.

APPLICATION OF THE POLYMERASE CHAIN REACTION TO DNA ENGINEERING

KERSTIN HAGEN-MANN

ESSC, Perkin-Elmer Cetus Analytical Instruments
Bahnhofstrasse 30, D-8011 Vaterstetten
(Federal Republic of Germany)

INTRODUCTION

A few years ago it was almost impossible to obtain information about a DNA sequence which was only present in a small
number of copies or a single molecule, respectively. The sequence of interest was either not detectable by conventional
methods or it was as contaminated with other DNAs as it was impossible to clone or sequence it. Since its publication in 1987
a method became one of the most powerful and versatile tools
for molecular biologists, clinicians, and other scientists: The
Polymerase Chain Reaction (PCR)[1]. This method enables us to obtain any kind of hidden, minute genetic information.

PCR is an *in vitro* method of nucleic acid synthesis by
which a particular segment of DNA can be specifically amplified
by repeated replications (Fig. 1). It involves two oligonucleotide primers that flank the DNA fragment of interest and repeated cycles of heat denaturation (1a) of the DNA, annealing
of the primers (1b) to their complementary sequences on the
target, and extension (1c) of the annealed primers with DNA
polymerase. The primers hybridize to opposite strands of the
target sequence and are oriented so that DNA synthesis by a DNA
polymerase proceeds across the region between the primers. The
extension products themselves are also complementary to the
primers, successive cycles essentially double the amount of the
target DNA synthesized in the previous cycle. The result is an
exponential accumulation of the specific target fragment,
theoretically 2^n fold the starting molecules, where n is the
number of cycles of amplifications performed.

PCR is sensitive to amplify single DNA molecules out of a
complex mixture of eukaryotic genomic sequences, individual
phage plaques or bacterial colonies. The amplified products can
be visualized as distinct bands on agarose gels. Enhancements
have fostered the development of numereous and diverse PCR
applications throughout the research community.

124

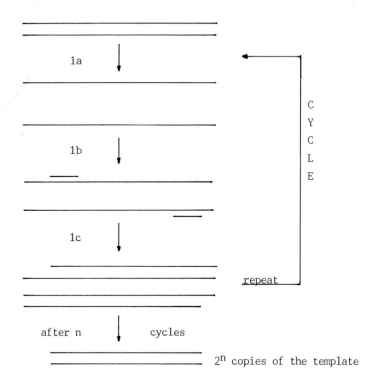

Fig. 1. Schematic description of Polymerase Chain Reaction
One cycle consists of three steps which are repeated n times
Step 1a: Heat denaturation of the dsDNA
Step 1b: Primer annealing on the denatured DNA single strands
Step 1c: Primer extension (replication) by DNA polymerase

PCR AND DNA MODIFICATION

PCR offers a tool for quick and easy modification of DNA
fragments. In comparison to recombinant DNA techniques, PCR
meets specific needs more efficiently. Sequence changes can
readily be made in the oligonucleotide primers chemically,
rather than by manipulating the DNA fragments with restriction
and ligation enzymes, e.g. PCR products can accept sequence
changes as 5'-"add-on" sequences (Fig. 2) to the primers. The
efficiency at which the product is modified is almost 100%.

Introduction of Restriction Site Sequences Into DNA Fragments[2]

The introduction of restriction sites into DNA fragments
which are produced by PCR is quite simple by attaching these
sequences to the 5'-ends of the primers[2].

125

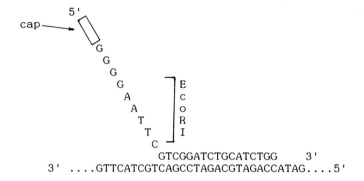

```
                          GTCGGATCTGCATCTGG      3'
        3' ....GTTCATCGTCAGCCTAGACGTAGACCATAG....5'
```

Fig.2. 5' "add-on" of a restriction site sequence[2]. A PCR
primer annealed to a target sequence with an added EcoRI
recognition sequence is shown. This addition does not affect
amplification and will be incorporated into the PCR product.
Often a cap is added to the 5' restriction site to ensure that
the restriction emzyme cleaves efficiently.

These sequences are mismatched to the template DNA but
normally this has little effect on the efficiency or specificity of the amplification; specificity is mostly dependent on
the match of the 3'-end of the primers. Strands whose replication is initiated by these "add-on" primers are themselves
copied and the added restriction site will dominate in the
growing population of PCR products. The principle of introduction of DNA alteration via PCR primers can be used to recombine
DNA sequences at any desired junction and to create sequence
alterations at any position. These sites facilitate insertion
of the product into cloning vectors, e.g. by the use of different sites at each end to create specific sticky ends for
ligation only in one direction into the vector.

Labeling of PCR Products

The creation of a labeled PCR product by 5'-end labeled
primers is the method of choice, rather than the incorporation
of labeled nucleotides during amplification. Applications of
end-labeling procedures include direct sequencing of the PCR
product or creation of DNA sequences suitable to be used in gel
retardation or protein detection assays. PCR is convenient
means of making DNA fragments available for *in vitro* protein
"footprint" analysis. This assay is independent of restriction
sites and requires only a few preparative steps.

Biotin or fluorescent lables can be attached to the 5'-end of a primer and remain unaffected in PCR performance. It has provided a means of non-isotopic detection with streptavidin or avidin-enzyme conjugates or of isolating one strand from the other by streptavidin-columns. Target-specific primers with different fluorophores can be used to distinguish the PCR products[3].

Overlap Extension Using PCR[4]

The Principle Reaction. Two DNA fragments with overlapping ends are generated by two complementary oligonucleotide primers used in PCR amplification (Fig.3A).

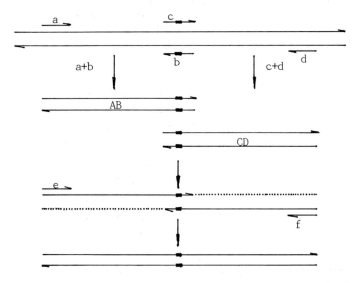

Fig. 3A. Schematic diagram of site-directed mutagenesis by overlap extension. The ds DNA and primers are represented by lines, the arrows indicate 5'-to 3' orientation. The site of mutagenesis is indicated by the black rectangle. Primer a and b produce fragment AB and primer c and d fragment CD. After denaturation the fragments reanneal at the overlap, are 3` extended by DNA polymerase and amplified by primers e and f.

In a subsequent fusion reaction in which these overlapping ends anneal the fragments are combined by 3'extension of the complementary strand. The 3'-ends serve as primers for replication and this resulting fusion product will serve as template for a second PCR.

Introduction of Point Mutations. Specific alterations in the nucleotide sequence can be introduced by incorporating nucleotide changes[7] into the overlapping primers(Fig. 3B).

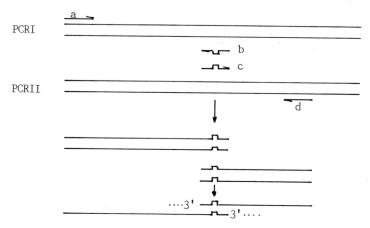

Fig. 3B. Site specific mutagenesis by introduction of point mutations. The primers c and d are mismatched to the target sequence in a single base. The mismatches lead to the same alteration in the two PCR products. PCR1 and PCR2 are performed separately, the products mixed after separation of excess primers, denatured and reannealed. The fusion product is formed and amplified as shown in Fig. 3A.

There is no requirement for restriction endonucleases or DNA ligase because the two PCR fragments generated have over-lapping ends that can effectively be fused in a subsequent PCR.

Insertion and Deletion Mutations. Insertions are possible because overlapping oligonucleotides only need to be comple-mentary to the template in the 3'region of the primer. The size of the insert is limited by the size of the primers (Fig.3C).

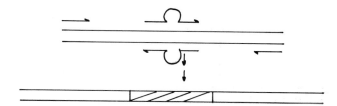

Fig. 3C. Use of complementary primers to place a small insertion into the ends of two overlapping fragments. The insertion is represented by the loop-out, but the insertion need not be flanked by the 5'-end by template complementary sequences as shown. The mutant fusion product is obtained and amplified as shown in Fig. 3A

Deletions are similarly obtained by designing primers such that their 3'-ends correspond to template sequences on one side of the deletion and the 5'-ends are complementary to template sequences on the other side of the deletion (Fig. 3D).

128

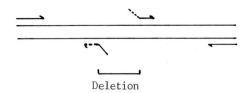

Deletion

Fig. 3D. Similar to the insertion overlapping primers can be used to create a deletion relative to the target sequence.

The Generation of Hybrid Genes. PCR fragments from unrelated sequences may be combined specifically through complementary overlaps created by 5' "add-on" sequences. These "add-on" sequences are complementary to the 3' region of the other primer. The resulting overlapping fragments are combined by chain extension of the recombinants via their recessed 3'-ends (Fig. 3F) which serve as primers for the 3' extension of the complementary strand. This method represents an important improvement over standard procedures because it is much faster and approaches almost 100% efficiency in the generation of mutant product.

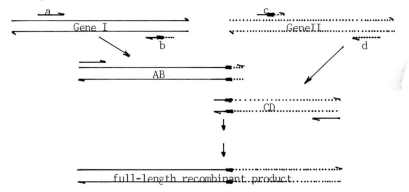

Fig. 3F. Two different genes are amplified. Gene I is shown as solid lines Gene II as dotted lines. The oligos a and b are primers for PCR I and amplify fragment AB, oligos c and d are primers for PCRII and amplify fragment CD. After amplification the PCR products are mixed, denatured, and reannealed. The end of one strand from each product is cabable of hybridizing with the complementary end from the other product. The full-length recombinant product is obtained by 3' extension.

Recombinant Circle PCR (RCPCR)[6]

Besides the overlap extension assay to introduce sequence alterations or to produce hybrid genes one method has been developed that uses PCR to generate products which can recombine to form circular DNA. An uncut plasmid with the insert of interest is amplified and mutated in two separate PCRs (Fig.4). The hybridization region of the 5'-ends of the PCR primers differs in both amplifications, the identical base pair is mutated. After the two separate PCR reactions the products are combined in one reaction tube, denatured, and reannealed. Due to the positions of the four PCR primers the single strands will form circles without restriction enzyme or ligation. The resulting plasmid can be transfected directly into competent cells. One of the most important advantages of this method is the fact that only 14 amplification cycles are required to obtain enough material for transfection. This reduces the frequency of misincorporations due to the lack of *Taq* DNA polymerase proofreading activity and many cycling steps.

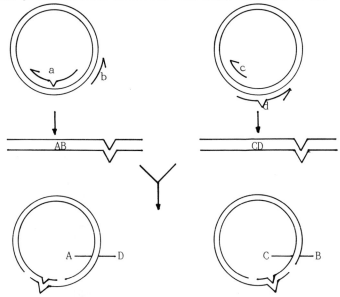

Fig. 4. Recombinant Circle PCR (RCPCR)[6]. An uncut plasmid is amplified in two separate PCRs. Primer a and b generate DNA strands A and B and primers c and d generate the strands C and D. Primers a and c are complementary to the same segment in the target, but oriented in opposite directions. Primers b and d are designed such that a strand from one PCR product can anneal to one strand of the other PCR product permitting the formation of circles of DNA.

130

CONCLUSION

The use of PCR primers that can easily be modified to introduce modifications and sequence alterations is a powerful
new tool in the study of genes and genetic engineering. Using
modified PCR primers is a simple way to label DNA fragments at
their 5'ends radioactively or non-radioactively.

PCR mutagenesis is applicable to any double-stranded DNA
sequence of suitable size to hybridize the upstream and downstream primer flanking a target site. The efficiency of PCR
mutagenesis is reasonable high (>80%) and it seems to be the
most facile and reliable way to create new DNA sequences by
production of overlapping fragments and subsequent combination.

REFERENCES

1. Mullis KB, Faloona FA (1987) Specific synthesis of DNA *in
    vitro* via a polymerase catalyzed chain reaction. In: Wu R,
    (ed) Methods in Enzymology 155:335-350

2. Higuchi R (1989) Using PCR to Engineer DNA. In: Erlich HA,
    (ed) PCR Technology:Principles and Applications for DNA
    Amplification. Stockton Press, New York, pp 61-70

3. Chehab FF, Kan YW (1989) Detection of specific DNA sequences
    by fluorescence amplification: A color complementation
    assay. Proc Natl Acad Sci USA 86:9178-9182

4. Ho SN, Hunt HD, Horton RM, Pullen JK, Pease LR (1989) Site-
    directed mutagenesis by overlap extension using the poly-
    merase chain reaction. Gene 77:51-59

5. Horton RM, Hunt HD, Ho SN, Pullen JK, Pease LR (1989)
    Engineering hybrid genes without the use of restriction
    enzymes: gene splicing by overlap extension. Gene 77:61-68

6. Jones DH, Howard BH (1990) A Rapid Method for Site-Specific
    Mutagenesis and Directional Subcloning by Using the Poly-
    merase Chain Reaction to Generate Recombinant Circles.
    BioTechniques 8:178-183

7. Kadowaki H, Kadowaki T, Wondisford FE, Taylor SI (1989) Use
    of polymerase chain reaction catalyzed by Taq DNA poly-
    merase for site-specific mutagenesis. Gene 76:161-166

DNA-PROTEIN INTERACTIONS

DNA-SEQUENCE INTERACTIONS

© 1991 Elsevier Science Publishers B.V. (Biomedical Division)
Site-Directed Mutagenesis and Protein Engineering
M.R. El-Gewely, editor.   133

CORRELATION BETWEEN THE HOMOLOGIES OF TTHHB8 AND TAQI ISOSCHIZOMERS AND INSERTION MUTANTS IN THE TAQI ENDONUCLEASE GENE

JOHN ZEBALA, ALAN MAYER and FRANCIS BARANY

Department of Microbiology, Hearst Microbiology Research Center, Cornell University Medical College, 1300 York Avenue, New York, NY 10021 (U.S.A.)

ABSTRACT

   The genes for TthHB8I restriction endonuclease and methylase (recognition sequence T↓CGA) were cloned in E. coli. A comparison of the restriction endonuclease DNA sequence to its isoschizomer, TaqI, reveals a 77% amino acid identity over the entire 263 amino acid length. Several codon insertion mutants in the TaqI endonuclease gene were constructed and activity characterized. The correlation between mutants with poor enzymatic activity and the regions of homology between TaqI and TthHB8I further suggests that these regions are involved in DNA recognition and/or catalytic cleavage.

INTRODUCTION

   Restriction endonucleases are unique DNA binding proteins that not only recognize a canonical DNA sequence, but also cleave with high specificity (20). These enzymes are ubiquitous in bacteria and appear to play a role in host defense from bacteriophage attack (7). A companion DNA methyltransferase recognizes the identical sequence and must always be present to protect the host DNA from endonuclease damage. This requirement is so absolute that the modification gene has always been found adjacent to the endonuclease gene in over 60 R-M systems that have been cloned (27).

   Important regions of DNA recognition proteins, such as the helix turn helix, leucine zipper, and zinc finger motifs have been discovered by comparing amino acid sequences (8, 11, 12). This approach has been highly successful with the RsrI and EcoRI isoschizomer endonucleases, where strong sequence homologies correlate with six regions implicated in DNA recognition by biochemical data and the three dimensional structure (22). However, no homologies were determined between the isoschizomeric HhaII and HinfI (4,17), or the approximately dozen other restriction endonucleases that have been sequenced (27). In contrast, sequence motifs specific to the methyltransferases have been observed (13,21,25).

We are interested in understanding the specificity of DNA recognition by the thermophilic restriction endonuclease, TaqI (recognition sequence T↓CGA, 14). The taqIR and taqIM genes have been cloned and sequenced (19). Isoschizomeric endonuclease and methyltransferase have previously been isolated from Thermus thermophilus strain HB8 (15,16). In this work, we clone the tthHB8IR and tthHB8IM genes, and compare the deduced amino acid sequence of the endonuclease to the taqIR gene. Using an in vivo genetic screen, we have isolated several two-codon insertion mutants which demonstrate a wide range of activities (1). Herein, we characterize additional insertion mutants, and correlate activities with sequence homologies.

RESULTS

(a) Cloning of the tthHB8IR and tthHB8IM genes. The TthHB8 restriction and modification genes were cloned by a biochemical selection protocol (23). DNA isolated from Thermus thermophilus strain HB8 was digested with BamHI, KpnI, SacI, HindIII, and PstI. Fragments were ligated into the appropriately cleaved and dephosphorylated vectors pTZ18R and pTZ19R (containing both a lac promoter and an f1 origin, M2), and ligation mixtures introduced into E. coli strain RRI. Transformants were selected by overnight growth at 32°C on plates containing ampicillin (50μg/ml). The resultant libraries (5,000 to 27,000 clones) were grown at 37°C under conditions which induce the lac promoter (1mM IPTG) to allow for in vivo self-methylation of any plasmids that acquired a functional methylase. Upon exhaustive digestion of total library plasmid DNA with TaqI (4,000 units for 2 hrs), only fully methylated plasmids remain intact, and these were rescued by reintroduction into E. coli. About half of such "biochemically selected" plasmids examined were resistant to TaqI, with the exception of plasmids isolated from the PstI library. The resultant plasmids; pFBTT2 (BamHI library), pFBTT3 (KpnI library), pFBTT4 (SacI library), pFBTT5 (HindIII library) contained overlapping inserts in the same orientation (See Fig. 1). Inverting the orientation of some of these inserts (plasmids pFBTT7, 8 and 9) resulted in plasmids that were only partially or weakly modified. We concluded that the tthHB8 methylase gene resided between the BamHI and HindIII sites, and required the upstream lac promoter for full modification of the host DNA.

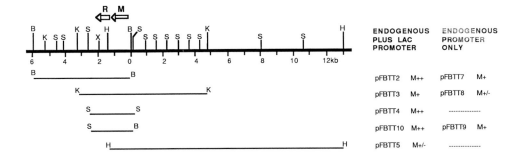

Fig. 1. Restriction map of Thermus thermophilus strain HB8 DNA in the region of the TthHB8 restriction (R) and modification (M) genes. Clones pFBTT2, 3, 4 and 5 were isolated from T. thermophilus DNA libraries (pTZ18R and 19R derivatives) as described in the text. These four clones contained the methylase gene downstream of the plasmid derived lac promoter (i.e. endogenous + lac promoter). To ascertain the strength of the T. thermophilus promoter in E. coli, inserts were inverted to avoid the lac promoter (endogenous promoter only). Cells were grown in the presence of IPTG (1mM) to induce the lac promoter. Methylation protection against digestion with TaqI restriction endonuclease: M++ = full protection, M+ = partial protection, M+/- = weak protection. Sites: B, BamHI; H, HindIII; K, KpnI; S, SacI; X, XmnI. Position of some SacI sites is approximate, but were used to confirm overlap and orientation of inserts. R = tthHB8 restriction endonuclease gene, M = tthHB8 methylase gene. Location and orientation of tthHB8 R and M genes was confirmed by sequencing. Dashed line indicates plasmid was not constructed.

(b) Sequencing of the tthHB8IR gene. The tthHB8IR and tthHB8IM genes were introduced into a pTZ vector containing unique 5' overhang (BamHI) and 3' overhang (PstI) sites in between the insert and universal sequencing primer site. This allowed for generation of directional deletions, from methylase towards endonuclease gene, using the processive action of ExoIII and trimming with S₁ Nuclease (6). Single stranded DNA prepared from these deletions (9) was sequenced using T7 DNA polymerase "Sequenase" as described (24). Analysis of the coding region for the endonuclease reveals a 77% amino acid identity between the tthHB8IR gene and the corrected taqIR gene sequence, throughout the entire 263 amino acid sequence (unpublished results). Nevertheless, 59% of the actual codons used differ, indicating that these two genes have undergone some evolutionary drift.

(c) Additional codon insertion mutants in the TaqI endonuclease gene. We expanded the sites for two-codon insertions in the TaqI endonuclease gene (1) by inserting special linkers into blunt end sites. The basic strategy is to randomly linearize a circular plasmid containing the TaqI endonuclease gene using a high frequency site enzyme. We used the "blunt end" enzymes: (i)

TABLE I. INSERTION MUTANTS IN TAQI ENDONUCLEASE

Plasmid (pFBIT)	Location	Number of Amino Acids	AMINO ACIDS INSERTED					Phenotype	Activity at 65 °C MHS Buffer	9LS Buffer
0.2	2	+2	*GLY*	ILE	PRO	Ser		H	10%	10%
0.5G	5 → 3	+5	Gln	GLY ILE	PRO	SER...	Gln	H	30%	30%
1	9	+2	Ala	ASP PRO	Leu			D	100%	100%
2	10	+2	Leu	GLY SER	Glu			D	100%	120%
3	22	+2	Lue	GLY SER	Glu			W	10%	150%
4	47	+2	Pro	ASP PRO	Asp			S	<0.1%	<0.1%
5	56	+2	Ala	ASP PRO	Leu			D	100%	100%
6	66	+2	Phe	GLY SER	Glu			W	10%	10%
7	77	+2	Arg	ILE ARG	Leu			W	100%	120%
7.1	78	+4 (fs)	Leu	ASP PRO	ASP PRO	PRO...		S	3%	3%
7.1f	78	+3	Leu	ASP ARG	SER	Pro	Leu	H	120%	100%
7.9	99	+2 (fs)	Leu	ASP PRO	ALA SER	ARG...		H	<0.1%	<0.1%
7.9	99	+3	Leu	ASP ARG	SER	Gly	Leu	S	<0.1%	<0.1%
8	101	+2	Leu	GLY SER	Glu			S	<0.1%	<0.1%
8.2	108	+2	Thr	GLY ILE	*PRO*			S	<0.3%	<0.3%
8.2G	108 → 80	+30 (fs)	Thr	GLY ILE	PRO LEU	GLY...		S	<0.3%	<0.3%
8.8	131	+2 (fs)	Glu	GLY ILE	LEU PRO	GLN...		S	<0.3%	<0.3%
8.8f	131	+3	Glu	GLY ILE	ASP *PRO*	Ala		S	N.C.	60%
8.9	131	+2 (fs)	Glu	GLY ILE	PRO SER	SER...		S	<0.3%	<0.3%
9	136	+2	Leu	GLY SER				S	N.C.	N.C.
9.2	153	+2,-1	Gly	SER Ile	Asp			D	30%	100%
9.5G	173 → 163	+12 (fs)	Arg	GLY SER	PRO PHE	ARG...		S	1%	20%
9.5Gf	173 163	+13 (fs)	Arg	GLY SER	ILE PRO	VAL...		S	<0.3%	<0.3%
10	181	+2	Ile	GLY SER	Glu			W	1%	3%
10.1	182	+2 (fs)	Glu	GLY SER	PHE PRO	THR...		S	<0.1%	<0.1%
10.1f	182	+3	Glu	GLY SER	ILE *LEU*	Pro		S	<0.1%	<0.1%
10.2G	183 → 174	+11	Ile	ARG ILE	GLU GLU	ARG...		W	3%	80%
11	193	+2	Leu	GLY SER	Glu			W	0.3%	10%
12	196	+2 (fs)	Lys	GLY SER	PRO ASN	GLY...		S	<0.3%	0.3%
12f	196	+3	Lys	GLY SER	ILE *PRO*	Lys		D	1%	5%
12.1	196 → 165	+33 (fs)	Lys	GLY SER	GLY SER	PRO...		W	0.3%	1%
12.1f	196 → 165	+34	Lys	GLY SER	ALA LYS	SER...		D	100%	100%
13G	224	+7	Asn	PRO GLU	PRO GLU	SER...		D	10%	60%
14	263	+38	Ala	GLY SER	ARG GLU	VAL...		(D)	100%	120%
14f	263	+3	Ala	GLY SER	ILE	Pro	End	D	100%	100%

'Location' is amino acid prior to insertion based on the revised nucleotide sequence of Taql endonuclease (unpublished results). 'Number of amino acid' refers to the number of amino acids inserted ('+') or deleted ('-'), and ('fs') means the change resulted in a frame shift. Bold amino acids in capitals are those which were inserted. Italic amino acids are those which have been changed as a result of the insert. "G" after the plasmid indicates the insertion results in internally duplicated domains known as "gemini" proteins. Phenotypes of patches grown overnight at 42°C under inducing conditions: D(dots), 3mm outline with a few persisters; H(hollow), 5mm, with less growth in center; S(small), 4mm; W(weak), 3-5mm less growth (Ref. 1). 'D' or 'W' morphology corresponded to high in vitro activity, while 'H' or 'S' showed little or no in vitro activity. Wild type enzyme gave 100% activity in both 9LS (10mM Tris HCl, pH9.0, 5mM MgCl$_2$) and MHS (10mM Tris HCl, pH8.0, 10mM MgCl$_2$, 100mM NaCl) buffer.

CviJI (recognition sequence PuG↓CPy), which cuts even more frequently in the presence of ATP (26) and (ii) MnlI (recognition sequence originally reported as CCTCN[7/7] ref.18) which we have now determined usually leaves a single base 5' overhang, i.e. CCTCN[6/7]. The second step is to ligate non-phosphorylated dodecamers (GGATCCGGATCC) to the 5' strand of each end; not only with randomly linearized plasmids but also to a HincII blunt end generated kanamycin cassette (2). Fragments and cassettes with such linker "tails" were purified from free linker, mixed at a one to two ratio, respectively, and ends phosphorylated with T4 kinase. A quick heat pulse (90ºC for 3 min) removes the non-covalently attached linker and allows for annealing of fragments to cassettes. Upon sealing the nicks with ligase, the resultant plasmids have kanamycin cassette inserts flanked by double BamHI sites. These plasmids were introduced into E. coli and screened for inserts within the TaqI endonuclease gene using a colony morphology assay when growing at 42ºC (1). Cassettes were excised, plasmids recircularized, reintroduced into E. coli, and resultant clones assayed for in vivo and in vitro activity as described (1). A summary of all insertion, gemini, frameshift, and deletion mutants characterized is provided in Table I.

DISCUSSION

TaqI endonuclease recognizes and cleaves its canonical sequence, TCGA, with complete fidelity under standard conditions by making a series of sequence specific interactions with the DNA duplex. Based on biochemical analysis of `star' site cleavages, we have hypothesized that sequence specific recognition is mediated by eight hydrogen bond contacts to the major groove (2). In order to help identify regions of TaqI which may be involved in sequence specific recognition and/or catalysis, we have cloned and sequenced the TthHB8I isoschizomer. Homologous regions between these proteins should suggest essential functional regions since they would be less likely to undergo evolutionary change. The two endonucleases showed 77% amino acid identity over the entire 263 amino acid length as well as eleven regions of homology containing five or more identical contiguous amino acids. The structural significance of these identical regions will have to await determination of the three dimensional structure of TaqI/DNA co-crystals.

In addition, we have made numerous codon insertions throughout the taqI endonuclease gene and examined their activity. The use of two-codon

138

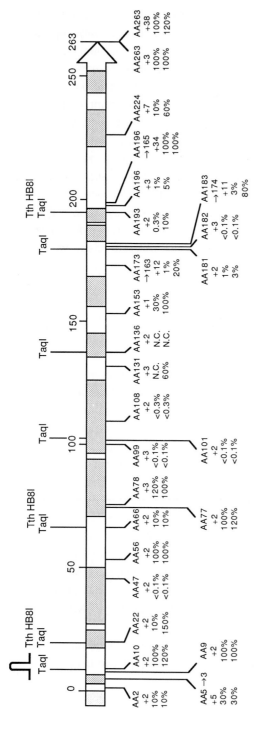

Fig. 2. Summary of amino acid identities of TaqI and TthHB8 isoschizomer, and in vitro endonuclease activity of two-codon insertions in the taqIR gene. Open arrow is a linear representation of the TaqI and TthHB8 endonuclease genes, both of which are 263 amino acids in length. Regions of identity greater or equal to five consecutive amino acids are shaded. The taqIR gene has seven TCGA sites (indicated by TaqI), while the tthHB8IR gene has only three such sites (indicated by TthHB8). The loop above the first TaqI site (at amino acid position 10) indicates the presence of a putative 14bp hairpin in the mRNA, which may play a role in regulating restriction endonuclease expression. Insertions listed in Table I are depicted below the arrow. AA (number) indicates the amino acid preceding the insertion. Percent activity compared to wild type is for MHS buffer (top line) and 9LS buffer (bottom line), respectively. Activities in excess of 100% may reflect increased yield of mutant protein. N.C. = enzyme activity produced nicked circles.

insertion mutagenesis as a potential tool for mapping functionally important regions in an enzyme has been discussed (Zebala & Barany this volume). The correlation between the mutants with poor activity and the regions of homology between the Taq I and TthHB8I isoschizomers further supports the notion that these regions are important for sequence specific DNA recognition and/or catalytic activity of the Taq I restriction endonuclease. Particularly intriguing are the very poor activity mutants at locations 47, 99, 101 and 108. These mutants are close to perfectly conserved arginines 41, 98, 104 and 107, respectively. We have previously identified three to four non-specific arginine interactions by NMR spectroscopy (5) and four phosphate contacts essential for cleavage from ethylation interference studies (28). It is interesting to speculate that these arginines might be involved in the required phosphate interactions. Site-directed mutagenesis of these residues will be required to investigate this hypothesis further.

ACKNOWLEDGEMENTS
We thank Antje Koller for expert technical assistance and Aneel Aggarwal, David Cowburn, Ira Schildkraut and Hamish Young for critical readings and helpful discussions. J.Z. and A.M. are both funded by MD-PhD fellowships. This work was supported by a grant from the National Institutes of Health (GM-41337-02) to F.B.

REFERENCES
1. Barany F (1987) Gene 56:13-27
2. Barany F (1988) Gene 63:149-165
3. Barany F (1988) DNA and Protein Engineering Techniques1:29-35
4. Chandrasegaran S, Lunnen KD, Smith HO, Wilson GG (1988) Gene 70:387-392
5. Glushka J, Barany F, Cowburn D (1989) Biochem Biophys Res Commun 164:88-93
6. Henikoff S (1987) Meth Enzymol 155:156-165
7. Kruger DH, Bickle TA (1983) Microbiol Rev 47:345-360
8. Landschulz WH, Johnson PF, McKnight SL (1988) Science 240:1759-1764
9. Mead DA, Szczesna-Skorupa E, Byron K (1986) Protein Engineering 1:67-74
10. Mead DA, Skorupa E, Byron K (1985) Nucl Acids Res 13:1103-1117
11. Miller J, McLachlan AD, Klug A (1985) EMBO J 4:1609-1614
12. Pabo CO, Sauer RT (1984) Annu Rev Biochem 53:293-321
13. Posfai J, Bhagwat AS, Roberts RJ (1988) Gene 74:261-265

14. Sato S, Hutchison CA, Harris JI (1977) Proc Natl Acad Sci USA 74:542-546
15. Sato S, Nakazawa K, Shinomiya T (1980) J. Biochem 88:737-747
16. Sato S, Shinomiya T (1978) J. Biochem 84:1319-1321
17. Schoner B, Kelly S, Smith HO (1983) Gene 24:227-236
18. Schildkraut I, Greenough L unpublished observations
19. Slatko BE, Benner JS, Jager-Quinton T, Moran LS, Simcox TG, Van Cott EM, Wilson GG (1987) Nucl Acids Res 15:9781-9796
20. Smith HO (1979) Science 205:455-466
21. Smith HO, Annau TM, Chandrasegaran S (1990) Proc Natl Acad Sci USA 87:826-830
22. Stephenson FH, Ballard BT, Boyer HW, Rosenberg JM, Greene PJ (1989) Gene 85:1-13
23. Szomolanyi E, Kiss A, Venetianer P (1980) Gene 10:219-225
24. Tabor S, Richardson CC (1987) Proc Natl Acad Sci USA 84:4767-4771
25. Tran-Betcke A, Behrens B, Noyer Weidner M, Trautner TA (1986) Gene 42:89-96
26. Xia Y, Burbank DE, Uher L, Rabussay D, Van Etten JL (1987) Nucl Acids Res 15:6075-6090
27. Wilson GG (1988) Gene 74:281-289
28. Zebala JA, Barany F, manuscript in preparation

© 1991 Elsevier Science Publishers B.V. (Biomedical Division)
Site-Directed Mutagenesis and Protein Engineering
M.R. El-Gewely, editor.

A GENETIC APPROACH TO STUDY THE STRUCTURE-FUNCTION RELATIONSHIP OF TRYPTOPHAN REPRESSOR

M. RAAFAT EL-GEWELY

Department of Biotechnology, Institute of Medical Biology, University of Tromsø, 9001 Tromsø, Norway.

ABSTRACT

Genetic regulation by *trp* repressor is one of the most thoroughly studied bacterial systems. It also provides a good model to study the structure-function relationship of proteins. In this article, the important aspects of this repressor are reviewed. Also, a new genetic approach to study structure-function relationship of multimeric proteins such as the *trp* repressor is discussed.

INTRODUCTION

The tryptophan repressor of *Escherichia coli* has been the subject of recent studies dealing with the structure-function relationship of proteins(1-8). This is due primarily to its simple primary structure, identified operator regions (9) and resolved crystal structure (10). The well defined *in vivo* and *in vitro* measurable activities resulting from its 1- binding to the operator region 2- binding to co-repressor made it possible to do such studies.

The repressor is encoded by the *trpR* gene which has been cloned and its nucleotide sequence was determined (9). The active repressor has two identical subunits made up of 107 amino acids and it binds two molecules of L- tryptophan as the corepressor. Active *trp* repressor binds its own operator, thus it is an autoregulated repressor (9). In addition, the active repressor binds two unlinked operons, *trp* operon (trpEDCBA) and *aroH* (1). *Trp* operon encodes the biosynthetic enzymes for tryptophan. The operator regions of these regulated operons have sequence similarities (9, 11).

Gene regulation by *trp* repressor:

The three operons that are regulated by *trp* repressor are regulated at different levels of repression *in vivo*, as recently reported (11).

Operon	Extent of repression
trp	300 fold
aroH	6 fold
trpR	3 fold

This range does not include the additional modulation due to attenuation (12). Using *in vivo* competition assay it was found that the repressor binds at approximately the same level to each of the three operators (11). Accordingly the repressor-operator affinity is not the explanation for the differential regulation of the *trp, aroH* and *trpR* operons observed *in vivo*. Klig et al., (11) proposed that this differential regulation observed *in vivo* may be due to the different relative locations of the three operators within their respective promoters.

In vitro studies:

Trp aporepressor has been purified to homogeneity (13, 14). This made it possible to preform several *in vitro* studies, in addition to the resolution of its structure by crystallography.

1- Repressor-operator binding assays:

Gel retardation experiments were recently made using purified repressor and cloned *trp* operator region (15), or synthetic operator oligonucleotide

143

(11, 16). It appears that the results of these groups are contradictory. Staake et al. (16) concluded that two *trp* repressor dimers, rather than one, bind to the *trp* operator. Previous results of Haydock et al. (17) would support the finding that two repressor dimers bind the operator since they demonstrated that half the operator region is still capable of binding *trp* repressor both *in vivo* and *in vitro*.

Analysis of *trp* repressor-operator interaction was made by filter binding (18). They suggested that the release of the corepressor may be the first step in dissociation of the repressor-operator complex.

2- Repressor-L-tryptophan binding assays:

The binding of L-tryptophan to the aporepressor was studied by equilibrium dialysis and flow dialysis (19, 20, 21). All these studies suggested that two independent molecules of ligand (L-tryptophan) bind to each native *trp* aporepressor dimer.

3- Heterodimer formation:

Purified *trp* aporepressors of the wild type and several mutants formed heterodimers upon treatment at 65°C for 3 min (21). Heterodimer formation was retarded by L-tryptophan or its analogs, suggesting that the indole moiety in the corepressor binding pocket increases the stability of the dimer.

TrpR structure:

Crystal structure of the *trp* repressor (10), *trp* aporepressor (22), *trp* pseudorepressor (23) *trp* repressor/operator complex (24) were resolved. The *trp* pseudorepressor is an inactive complex formed when the *trp* aporepressor binds indole analogues of L-tryptophan. Indole analogues unlike tryptophan confer no affinity for the *trp* operator. These crystal structure studies have shown that the dimeric *trp* repressor has a remarkable subunit

		Helix								Turn			Helix									
		1	2	3	4	5	6	7	8	9	10	11	12	13	14	15	16	17	18	19	20	
Trp Rep	68 –	Gln	Arg	Glu	**Leu**	**Lys**	Asn	Glu	**Leu**	**Gly**	Ala	Gly	Ile	Ala	Thr	**Ile**	Thr	Arg	**Gly**	Ser	Asn	– 87
Lac Rep	6 –	Leu	Thr	Asp	**Val**	**Ala**	Arg	Leu	**Ala**	**Gly**	Val	Ser	Tyr	Gln	Thr	**Val**	Ser	Arg	**Val**	Val	Asn	– 25
λ Rep	33 –	Gln	Glu	Ser	**Val**	**Ala**	Asp	Lys	**Met**	**Gly**	Met	Gly	Gln	Ser	Gly	**Val**	Gly	Ala	**Leu**	Phe	Asn	– 52
CAP	169 –	Arg	Gln	Glu	**Ile**	**Gly**	Glu	Ile	**Val**	**Gly**	Cys	Ser	Arg	Glu	Thr	**Val**	Gly	Arg	**Ile**	Leu	Lys	– 188
434R	17 –	Gln	Ala	Glu	**Leu**	**Ala**	Gln	Lys	**Val**	**Gly**	Thr	Thr	Gln	Gln	Ser	**Ile**	Glu	Gln	**Leu**	Glu	Asn	– 36
P22 Rep	21 –	Gln	Ala	Ala	**Leu**	**Gly**	Lys	Met	**Val**	**Gly**	Val	Ser	Asn	Val	Ala	**Ile**	Ser	Gln	**Trp**	Glu	Arg	– 40
P22 Cro	13 –	Gln	Arg	Ala	**Val**	**Ala**	Lys	Ala	**Leu**	**Gly**	Ile	Ser	Asp	Ala	Ala	**Val**	Ser	Gln	**Trp**	Lys	Glu	– 32
λ cII	26 –	Thr	Glu	Lys	**Thr**	**Ala**	Glu	Ala	**Val**	**Gly**	Val	Asp	Lys	Ser	Gln	**Ile**	Ser	Arg	**Trp**	Lys	Arg	– 45
P22 cI	1 –	Gln	Arg	Lys	**Val**	**Ala**	Asp	Ala	**Leu**	**Gly**	Ile	Asn	Glu	Ser	Gln	**Ile**	Ser	Arg	**Trp**	Lys	Gly	– 19
Mat α*	117 –	Lys	Glu	Glu	**Val**	**Ala**	Lys	Lys	**Cys**	**Gly**	Ile	Thr	Pro	Leu	Gln	**Val**	Arg	Val	**Trp**	Cys	Asp	– 136
Antp*	31 –	Arg	Ile	Glu	**Ile**	**Ala**	His	Ala	**Leu**	**Cys**	Leu	Thr	Glu	Arg	Gln	**Ile**	Lys	Ile	**Trp**	Phe	Gln	– 50

Fig. 1. The helix-turn-helix motif (HTH) of *trp* repressor as compared to other prokaryotic HTH and eucaryotic* HTH elements (25,26,27). Amino acids that are characteristic for the position are shown in **Bold**. The numbers of amino acid residues of each protein is indicating the beginning and the end of the HTH element.

interface in which five of six helices of each subunit are interlinked. The inactive aporepressor must bind two molecules of L-tryptophan to form the active repressor. The most striking finding is that the crystal structure shows no direct hydrogen-bonded or non-polar contacts with the bases that could explain in simple chemical terms the repressor's specificity in binding to the operator region. However recent results (16) have demonstrated that the synthetic oligonucleotide that was used as the operator sequence in the crystal structure study of the *trp* repressor complex did not bind to the repressor in gel retardation experiments. It remains to be seen if the chemical nature of repressor-DNA interaction will be further elucidated in crystal studies using a synthetic operator oligonucleotide that has been tested by gel retardation experiments.

TrpR-DNA interaction:

1- The DNA binding Motif:

The *trp* repressor follows the helix-turn-helix motif (HTH) as reviewed by Pabo and Sauer (25) and Harrison and Aggarwal (26).

L- tryptophan binds between the central core and the HTH, in a pocket formed by glycine 85 and arginine 84 from one subunit of the repressor and threonine 44 from the other subunit. Glycine 85 in tryptophan repressor, HTH position 18, is unique. In other prokaryotic repressors with the HTH motif, large hydrophobic residues are common (10). Several prokaryotic repressors contain tryptophan at position 18 in the HTH motif (25, 26) as indicated in Figure 1. It is interesting to note that in all studied homeodomain amino acid sequence, position 18 in the HTH motif is always occupied by L-tryptophan (27, 28). The helix2-helix3 of *Ant* (*Antpannapedia*) homeodomain is listed in Fig. 1, as an example of homeodomain.

We replaced glycine 85 (HTH position 18) with tryptophan. This mutant GW85 gave a repressor with reduced repressor activity. However this activity was independent of L-tryptophan in the media. We are reporting the result of this study elsewhere (29).

2- Operator regions:

The nucleotide sequence of the three operator regions of the genes regulated by trpR shows similarity, but they are not identical (11). The sequence of these operator regions is as indicated below:

```
                                center of symmetry
                                        ↓
trp operator (-22 to -2):       TCGAACTAGTT.AACTAGTACG

aroH operator (-47 to -28):      TGTACTAGAG.AACTAGTGCA

trpR operator (-11 to + 10):    TCGTACTCTTT.AGCGAGTACA

Common symmetry:                --G-ACT----.----AGT-C-

Consensus Sequence:             TCGTACTAGTT.AACTAGTACA
```

In a recent publication, Staacke et al. (16), based on a proposed model for protein-DNA recognition (30, 31), has suggested new binding sites for the trp repressor. In this model the trp operator has two centers of symmetry allowing the operator to bind two repressor molecules. According to their model the proposed sequence for operator region should be as below:

```
Classical consensus sequence:  TCGTACTAGTT.AACTAGTACA
                               AGCATGATCAA.TTGATCATGT

New model:                     CTATCGTACTAGTT.AACTAGTACGATAG
                               GATAGCATGATCAA.TTGATCATGCTATC
```

General genetic approaches to study structure-function relationship:

1- Second site reversions

2- Negative complementation

3- Intragenic complementation

1- Second-site reversion analysis:

   In this classical technique, the activity of a mutant protein can in some cases be rescued by amino acid substitutions that does not revert back to the same wild type sequence. This technique is used to provide useful information about the relationship between the primary structure of a given protein and its tertiary structure (32, 33, 34). This approach was recently utilized to study 5 different *trp* repressor mutants (3). They reported that most of the second-site revertant repressors were found to have the same amino acid substitutions detected previously as superrepressor mutants (1). These superrepressor changes at different sites in the protein can act globally to increase the activity of the primary mutant repressors.

2- Negative complementation:

   This type of analysis is only operative in proteins with more than one polypeptide and in cases of protein-protein interaction. If a protein is multimeric, a negative-complementing (trans-dominant) mutant will produce a protein capable of interacting with wild-type polypeptide chains if both forms of the genes are placed together *in vivo*. This will result in an inhibitory effect to the final activity/function *in vivo* by forming a multimeric protein with impaired function. This approach has been utilized to study LacI^{-d} mutants (35, 36) and several *trp* repressor mutants (1). The negative-complementing *trp* mutants were all missense or nonsense affecting the different domains of the *trp* repressor. Negative complementing or "dominant

148

negative" mutants were proposed as a general way to assign functions to different genes in cases of protein-protein interaction (37).

3- Intragenic complementation:

As in negative complementation, intragenic complementation is operative in proteins with more than one polypeptide chain. In intragenic complementation, some mutants in the same cistron/gene can complement each other restoring the wild type activity or approaching it *in vivo*. This subject of intragenic complementation was studied in many genetic systems (38).

This approach is suitable to study the structure-function relationship of multimeric proteins *in vivo*. We are now studying intragenic complementation of *trp* repressor mutants affecting the L-tryptophan binding or DNA binding in the operator region. The effect of amino acid charge or size at different sites of the interacting mutant polypeptides on the activity of the heterodimers formed *in vivo* is studied. The DNA binding activity of repressors that are the result of an intragenic complementation of defective mutants is also being studied. The details of this study will be reported elsewhere (29).

ACKNOWLEDGMENTS

I would like to express my thanks to Dale Oxender and Charles Yanofsky for their encouragement and for making the strains available to us. This work was supported by Norwegian Society for Science and Humanities (NAVF).

REFERENCES

1. Kelley, R.L., Yanofsky, C. (1985) Mutational studies with the trp repressor of Escherichia coli support the helix-turn-helix model of repressor recognition of operator DNA. Proc. Natl. Acad. Sci. USA 82: 483-487.

2. Bass, S., Sorrells, V., Youderian, P. (1988) Mutant *trp* repressors with new DNA-binding specificities. Science 242: 240-245.

3. Klig, L.S., Oxender, D.L., Yanofsky, C. (1988) Second-site revertants of *Escherichia coli trp* repressor mutants. Genetics 120: 651-655.

4. Chou, W.-Y., Bieber, C., Matthews, K.S. (1989) Tryptophan and 8-anilino-1-naphthalenesulfonate compete for binding to trp repressor. J. Biol. Chem. 264: 18309-18313.

5. Chou, W.-Y., Matthews, K.S. (1989) Serine to cysteine mutations in trprepressor protein alter tryptophan and operator binding. J. Biol. Chem. 264: 18314-18319.

6. Gittelman, M.S., Matthews, C.R. (1990) Folding and stability of *trp* aporepressor from *Escherichia coli*. Biochem. 29: 7011-7020.

7. He, J.-J., and Matthews, K. S. (1990) Effect of amino acid alterations in the tryptophan-binding site of the *trp* repressor. J. Biol. Chem. 265: 731-737.

8. Hurlburt, B.K., and Yanofsky, C. (1990) Enhanced operator binding by *trp* superrepressors of *Escherichia coli*. J. Biol. Chem. 265: 7853-7858.

9. Gunsalus, R.P., Yanofsky, C. (1980) Nucleotide sequence and expression of *Escherichia coli trpR*, the structural gene for the trp repressor. Proc. Natl. Acad. Sci. USA 77: 7117-7121.

10. Schevitz, R.W., Otwinowski, Z., Joachimiak, A., Lawson, C.L.,

Sigler, P.B. (1985) The three-dimensional structure of *trp* repressor. Nature 317: 782-786.

11. Klig, L.S., Carey, J., Yanofsky, C. (1988) *trp* Repressor interactions with the *trp*, *aroH*, and *trpR* operators. Comparison of repressor binding *in vitro* and repression *in vivo*. J. Mol. Biol. 202: 769-777.

12. Yanofsky, C. (1984) Comparison of regulatory and structural regions of genes of tryptophan metabolism. Mol. Biol. Evol. 1: 143-161.

13. Joachimiak, A., Kelley, R.L., Gunsalus, R.P., Yanofsky, C., Sigler, P.B. (1983) Purification and characterization of trp aporepressor. Proc. Natl. Acad. Sci. USA 80: 668-672.

14. Paluh, J.L., Yanofsky, C. (1986) High level production and rapid purification of the *E. coli trp* repressor. Nucleic Acids Res. 14: 7851-7860.

15. Carey, J. (1988) Gel retardation at low pH resolves *trp* repressor-DNA complexes for quantitative study. Proc. Natl. Acad. Sci. USA 85: 975-979.

16. Staacke, D., Walter, B., Kisters-Woike, B., Wilcken-Bergmenn, B. v., Müller-Hill, B. (1990) How *trp* repressor binds to its operator. EMBO J. 9: 1963-1967.

17. Haydock, P.V., Bogosian, G., Brechling, K., Somerville, R.L. (1983) Studies on the interaction of trp holorepressor with several operators. J. Mol. Biol. 170: 1019-1030.

18. Klig, L.S., Crawford, I.P., Yanofsky, C. (1987) Analysis of *trp* repressor-operator interaction by filter binding. Nucleic Acids Res. 15: 5339-5351.

19. Arvidson, D.N., Bruce, C., Gunsalus, R.P. (1986) Interaction *Escherichia coli trp* of the aporepressor with its ligand, L-tryptophan. J. Biol. Chem. 261: 238-243.

20. Marmorstein, R.Q., Joachimiak, A., Sprinzl, M., Singler, P.B. (1987) The structural basis for the interaction between L-tryptophan and *Escherichia coli trp* aporepressor. J. Biol. Chem. 262: 4922-4927.

21. Graddis, T.M., Klig, L.S., Yanofsky, C., Oxender, D. (1990) Formation of heterodimers between wild type and mutant *trp* aporepressor polypeptides of *Escherichia coli*. Genetics 120: 651-655.

22. Zhang, R.G., Joachimiak, A., Lawson, C.L., Otwinowski, Z., Sigler, P.B. (1987) The crystal structure of trp aporepressor at 1.8A shows how binding tryptophan enhances DNA affinity. Nature 327: 591-597.

23. Lawson, C.L., Sigler, P.B. (1988) The structure of trp pseudorepressor at 1.65A shows why indole propionate acts as a trp "inducer". Nature 333: 869-871.

24. Otwinowski, Z., Schevitz, R.W., Zhang, R.-G., Lawson, C.L., Joachimiak, A., Marmorstein, R.Q., Luisi, B.F., and Sigler, P.B. (1988). Crystal structure of *trp* repressor/operator complex at atomic resolution. Nature 335: 321-329.

152

25. Pabo, C.O., and Sauer, R.T. (1984). Protein-DNA recognition. Annu. Rev. Biochem. 53: 293-321.

26. Harrison, S.C., Aggarwal, A.K. (1990) DNA recognition by proteins with the helix-turn-helix motif. Annu. Rev. Biochem. 59: 933-969.

27. Scott, M.P., Tamkun, J.W., Hartzell III, G.W. (1989) The structure and function of the homeodomain. Biochimica et Biophysica Acta 989: 25-48.

28. Gehering, W.J., Muller, M. Affolter, M., Percival-Smith, A., Billeter, M., Qian, Y:Q., Otting, G., and Wutherich, K. (1990) The structure of the homeodomain and its functional implications. TIG 6: 323-328.

29. Storbakk, N., Oxender, D.L., Yanofsky, C., and El-Gewely, M.R. (1990) Intragenic complementation in *trp* repressor of *Escherichia coli*. Manuscript in preparation.

30. Lehming, N., Sartorius, J., Kisters-Woike, B., Wilcken-Bergmann, B. v., Müller-Hill, B. (1990) Mutant *lac* repressors with new specificities hint at rules for protein - DNA recognition. EMBO J. 9: 615-621.

31. Lehming, N., Sartorius, J., Kisters-Woike, B., Wilcken-Bergmann, B. v., Müller-Hill, B. (1990) Rules for Protein/DNA-recognition. This Proceedings.

32. Helinsky, D.R. and Yanofsky, C. (1963) A genetic and biochemical analysis of second-site reversion. J. Biol. Chem. 238: 1043-1048.

33. Hecht, M.H., and Sauer, R.T. (1985) Phage lambda repressor revertants: amino acid substitutions that restore activity to mutant proteins. J. Mol. Biol. 186: 53-63.

34. Nelson, H.C.M., and Sauer, R.T. (1985) Lambda repressor mutations that increase the affinity and specificity of operator binding. Cell 42: 549-558.

35. Adler, K., Beyreuther, K., Fanning, E., Geisler, N., Gronenborn, B., Klemm, A., Muller-Hill,B., Pfahl, M., and Schmitz, A. (1972) How *lac* repressor binds to DNA. Nature 237: 322-327.

36. Miller, J.H. (1980) in *The Operon* eds., Miller, J.H., Rezenkoff, W.S. (Cold Spring Harbor, NY), 31-88.

37. Herskowitz, I. (1987) Functional inactivation of genes by dominant negative mutations. Nature 329: 219-222.

38. Fincham, J.R.S. (1966) Genetic Complementation. W.A. Benjamin, Inc., New York.

© 1991 Elsevier Science Publishers B.V. (Biomedical Division)
Site-Directed Mutagenesis and Protein Engineering
M.R. El-Gewely, editor.

RULES FOR PROTEIN/DNA-RECOGNITION

NORBERT LEHMING*, JÜRGEN SARTORIUS, BRIGITTE KISTERS-WOIKE,
BRIGITTE VON WILCKEN-BERGMANN, and BENNO MÜLLER-HILL

Institut für Genetik der Universität zu Köln, Weyertal 121, D-5000
Köln 41 (FRG)

*present address: Department of Biochemistry and Molecular Biology,
Harvard University, 7 Divinity Avenue, Cambridge MA 02138 (USA)

INTRODUCTION

Protein/DNA-recognition plays a central role in transcriptional
regulation of gene expression. Several different motifs for DNA-
binding have been identified, such as the helix-turn-helix (HTH)
motif (1), the zinc finger (2), the basic domain of leucine zipper
proteins (3), the extended ß-sheet (4), and the helix-loop-helix
(HLH) motif (5). We will only present data for the class of HTH-
motif containing proteins. It remains to be seen, if the results
can be transferred to other systems.

MATERIALS AND METHODS

Media, plasmids, strains, and general methods were as described
in (6-9).

RESULTS

We have set up a system which allows fast testing of protein/DNA-
interactions (6). The system consists of two compatible plasmids.
One plasmid carries a semisynthetic lac I gene, an amp resistance
gene, and the origin of replication from pBR322. The other plasmid
carries the lac Z gene with a unique XbaI-site instead of wt lac
operator, a tet resistance gene and the origin of replication from
pACYC 184. The two origins are compatible. Any lac operator variant
can be cloned into the unique XbaI-site. Any amino acid in the Lac
repressor headpiec can be replaced by exchanging small pieces of
synthetic double stranded DNA. The repression value (Specific
activity of ß-galactosidase in the absence of Lac repressor (mutant)
divided by the specific activity of ß-galactosidase in the presence
of Lac repressor (mutant)) yields a direct value for the affinity of

this protein to the operator cloned into the XbaI-site. In vitro
studies have been carried out to support the in vivo data (6-9).
Table I shows the repression values of wt Lac repressor and several
Lac repressor mutants for the ideal lac operator and a set of single
symmetric lac operator exchanges.

TABLE I

REPRESSION OF LAC OPERATOR VARIANTS BY LAC REPRESSOR MUTANTS

Repression values are given single symmetric exchanges of lac
operator in the presence of wt Lac repressor and of Lac repressor
mutants. V1: tyrosine 17 of Lac repressor, which is the first
amino acid of the recognition helix, has been replaced by a valine.
A2: glutamine 18 has been replaced by an alanine. V1A2: both amino
have been replaced. N6: arginine 22 has been replaced by an asp.

pWB	operator		wt	V1	A2	V1A2	N6
310	AATTGTGAGC	GCTCACAATT	≥200	64	20	24	1
332	AATTGTGCGC	GCGCACAATT	2	2	1	1	1
333	G	C	2	2	1	1	1
334	T	A	2	1	1	1	1
341	AATTGTAAGC	GCTTACAATT	5	2	50	≥200	1
342	C	G	6	2	1	2	1
344	T	A	4	18	1	18	1
351	AATTGAGAGC	GCTCTCAATT	16	15	2	25	1
352	C	G	10	8	2	3	1
353	G	C	9	12	1	20	1
361	AATTATGAGC	GCTCATAATT	3	2	1	1	1
362	C	G	2	1	1	1	1
364	T	A	3	2	1	1	100

As can be seen from the repression values, position 6 of Lac repressor recognition helix binds to base pair number 6 of lac operator, numbered from the center of symmetry. The amino acids in
position 1 and 2 of Lac repressor recognition helix bind to base
pairs number 4 and 5. Amino acid number 6 acts clearly independently
from amino acids 1 and 2. But in what way do these two interact?
To address this question, we tested all 20 amino acids in position 1
and 2 with all 16 possible lac operator double variants in position
4 and 5. We found out, that also the amino acids 1 and 2 act independently. We concluded this from an analysis as given in Table
II. The amino acids do not change their individual affinity to
the given operator. To do the calculation for all repression values
at the same time, we split the repression value into two factors.
Each factor is supposed to represent the binding energy of the
single amino acid.

TABLE II

SPLITTING THE REPRESSION VALUES INTO FACTORS

Calculation of 13 repression values. The factors for the four amino acids in position 1 and the four amino acids in position 2 are calculated in a way to minimize the difference between the theoretical (given in brackets) and the measured repression value. The factors are given beneath the symbols for the amino acids. Note that thee repression values have not been measured, but can be calculated from the other ones. Whenever checked, the predictions turned out right. The factor 1 is given arbitrarily to G1.

344	G 1 (1)	E 1 (0.5)	P 1 (2)	S 1 (3)
A 2 (8)	8 (8)	4 (4)	– (16)	17 (24)
M 2 (60)	60 (60)	30 (30)	100 (120)	– (180)
Q 2 (30)	– (30)	10 (15)	50 (60)	100 (90)
K 2 (1)	2 (1)	1 (1)	2 (2)	2 (3)

DISCUSSION

We have shown that in the case of Lac repressor distinct amino acids prefer distinct bases in distinct positions. A given amino acid binds different bases if it is in different positions. This shows that there is no general code for recognizing of base pairs by amino acids. But if we compare the specificity of a given amino acid in a given position to a given base pair in different repressor operator systems, such as the gal system, we find that the specifi city of the amino acid remains (9). Thus we call the specificities found in the lac system : "Rules".

REFERENCES

1. Anderson et al. (1981) Nature 290:754-758

2. Miller et al. (1985) EMBO J 4:1609-1614

Landschulz et al. (1988) Science 240:1759-1764

4. Knight et al. (1989) J Biol Chem 264:3639-3642

5. Murre et al. (1989) Cell 56:777-783

6. Lehming et al. (1987) EMBO J 6:3145-3153

7. Lehming et al. (1988) Proc Natl Acad Sci USA 85:7947-7951
8. Sartorius et al. (1989) EMBO J 8, 1265-1270
9. Lehming et al. (1990) EMBO J 9:615-621

SITE-DIRECTED MUTAGENESIS

© 1991 Elsevier Science Publishers B.V. (Biomedical Division)
Site-Directed Mutagenesis and Protein Engineering
M.R. El-Gewely, editor.

OLIGONUCLEOTIDE AND MULTI SITE-DIRECTED MUTAGENESIS

M. RAAFAT EL-GEWELY

Department of Biotechnology, Institute of Medical Biology, University of Tromsø, 9001 Tromsø, Norway.

INTRODUCTION

The techniques of site-directed mutagenesis are rather simple, and they are continuously getting simpler. But that does not mean that it is an easy subject to review fully, because of the continuous improvements and innovations. Mutagenesis has always been a powerful tool in genetic research. With the progress in molecular biology and the development of recombinant DNA technology such an approach became more precise. The logical evolution of the methods in molecular biology that enabled researchers to isolate, clone and sequence genes, led to the development of needed techniques that introduce defined mutations in the cloned genes *in vitro*. This subject was reviewed by Smith (1, 2) and Knowles (3).

Site-directed mutagenesis is usually performed for the following reasons: 1- To study structure-function relationships of proteins 2- To do gene reconstruction experiments for heterologous gene expression experiments (4, 5). 3- Synthetic mutations for human diseases. This new approach will enable scientists to make mutations *in vitro*, mimicking mutations discovered only by protein sequence studies. This approach is simpler than the option of constructing a cDNA library of the appropriate tissues, where the gene is expressed, of the new patient (tissues might not be available), and screening for the corresponding gene. This of course with assumption that such cDNA library exist, or it can be constructed one time only and all the possible mutations could then be made *in vitro*. Engineering such synthetic mutations would eventually enable scientists to establish animal models, for example, carrying specific mutation(s). Also, physical, biochemical characterization of the mutant protein could be studied after its heterologous expression in an appropriate system.

Following the introduction of desired mutations, an expression of the mutated gene is required to elaborate on its function/phenotype *in vivo* or *in vitro*. Most of the gene expression of mutated proteins for *in vivo* studies, have been done in *Escherichia coli* and recently in yeast and mammalian cells.

Classes of *in vitro* mutagenesis methods:

Several approaches and methods for site-directed mutagenesis have been described. These methods can basically be summarized as in Table I.

TABLE I

IN VITRO MUTAGENESIS

Summary of the different techniques used in *in vitro* mutagenesis

I. Random	II. None Random	
	a. Complete synthesis of gene/fragment	b. Oligonucleotide- directed mutagenesis
1. mutagenesis of DNA	1. organic synthesis	1. substitution
2. linker mutagenesis	2. PCR	2. deletion
3. PCR		3. insertion
		4. site/region saturation
		5. multisite/gene reconstruction
		6. incorporation of unnatural amino acids
		7. PCR

I. Random mutagenesis:

Random mutagenesis methods aim at the introduction of point mutations throughout the molecule and this can be accomplished in different ways. These methods offer the advantage of mapping the functional domains of the protein without bias.

I.a. Direct mutagenesis of DNA:

I.a.1. Chemical methods:

In these methods DNA is treated *in vitro* with reagents that change the specificity of base pairing leading to base substitution after *in vivo* or *in vitro* replication of the treated DNA molecules. Several chemicals have been used, for example, sodium bisulfite which reacts preferably to single stranded DNA (6, 7), Hydrazine (8), Hydroxylamine (9) and methoxylamine (10).

I.a.2. Misincorporation of deoxynucleotides:

Using the DNA understudy as a template and using only three of the four deoxynucleotide triphosphates, together with an appropriate DNA polymerase, enforces misincorporation of bases *in vitro*. This leads to random mutation in the DNA molecules after their replication *in vivo* (11).

I. b. Linker mutagenesis:

It is done by introducing at random oligonucleotide sequence coding, for example, for two amino acids (12, 13). In this method single stranded hexameric linkers are inserted into a plasmid which has been linearized at cohesive-ends. The Plasmids containing linker insertions are enriched by using biochemical or biological selection. This method offers the advantage of mapping the protein functional domains without introducing any possible frame-shift mutants.

II. Non random methods for *in vitro* mutagenesis:

II. a. Complete synthesis of the gene/fragment :

II. a. 1. organic synthesis of DNA:

A complete chemical synthesis of a gene or a small segment of a gene containing the desired mutation(s) replaces a natural segment or a whole gene. Chemical synthesis of DNA usually yields a large amount of DNA and the codon usage could be adjusted to suit a heterologous gene expression situation (14).

II. a. 2. PCR:

Polymerase chain reaction technique (15, 16) allows the enzymatic amplification of any DNA segment using two oligonucleotide primers and a thermostable DNA polymerase such as *Taq* polymerase. This technique is opening new frontiers in molecular biology and is fast replacing cloning as a strategy. This technique has been recently used to introduce site-specific mutations (17-20). The use of PCR in site-directed mutagenesis is reviewed by Kirstin Hagen-Mann in this publication (21). A simple method to make gene reconstructions during the cloning of genes by PCR is presented by our group (5). The amplification of the gene is done by two primers flanking the region of interest (in opposite direction) and *Taq* polymerase. One of the primers usually contain the required mutation/mismatch. Inverse PCR, using back-to back primers, was also used to introduce site specific mutations (22).

II. b. Oligonucleotide-directed mutagenesis:

Oligonucleotide-site directed mutagenesis are the most commonly used. Several strategies have been used (1, 2, 23, 24, 25, 4). One of the most common problems of oligonucleotide-directed mutagenesis is the low frequency of mutants. To overcome this problem, two basic approaches have been developed to enrich for the resulting mutants,   the biochemical approach (23, 26, 27) and the genetic approach (24, 25). The biochemical method of selecting mutants is presented by David Olsen in this symposium (27).

In both approaches for the enrichment of mutants usually a single-stranded DNA template, e.g. "+" strand of M13 DNA, recombinant for the desired gene/DNA fragment is annealed to a mutagenic primer (synthetic oligonucleotide containing the desired mutant). The primer is then enzymatically extended *in vitro* producing heteroduplex. After introduction to *Escherichia coli* the heteroduplex is replicated *in vivo* and then resolved to mutant and parental molecules (non-mutated molecules). Verification of the mutant sequence is done by DNA sequencing using Sanger's method (28).

In oligonucleotide mutagenesis, progress lies mainly in improving the efficiency of the enzymatic extension and the biological or biochemical elimination of the parental genotype.

We have used the genetic selection method developed by (24, 25) in combination with a processive polymerase (T7 Polymerase) (29) to introduce simple site- and multisite-directed mutagenesis. When we used mutagenic oligonucleotides containing base substitutions and major deletions or insertions simultaneously, only modified T7 polymerase produced the designed mutants (4).

An *Escherichia coli* strain with double mutants is required for this selection, "RZ1032" (25). These mutants are, 1. dut- (it codes for dUTP nucleotide hydrolase : dUTP $\longrightarrow$ dUMP + PPi). Such a mutant will have high levels of dUTP, thus the synthesized DNA will have "U" in place of "T". 2. The other mutant is ung- (uracil N-glucosidase). This enzyme removes uracil from DNA, thus such a mutant is needed to keep the DNA intact. Maximum mutation frequency was reached within 15 minutes with high efficiency.

166

The mutated gene is subcloned into the appropriate expression vector for *in vivo* studies or for heterologous gene expression. We have used the technique successfully for both purposes using *E. coli* trp repressor and pGH (4).

The discussed technique of site- and multisite-directed mutagenesis could also be applicable for making *region and site saturation* mutagenesis in addition to the *substitution, deletion, insertion, and gene reconstruction experiments. Site and region saturation mutagenesis* are based on the utilization of synthetic oligonucleotides that have degeneracy at any number of positions. A degenerate oligonucleotide at any position will have the three non-wild-type nucleotides at same position during the synthesis of the oligonucleotide. This results in a mixture of related molecules (30, 31).

Techniques that can introduce unnatural amino acids into specific sites in the protein using chemically acylated suppressor tRNAs were developed (32). Site-directed mutagenesis methods using double stranded vectors were reported (26, 33).

ACKNOWLEDGEMENTS
This work was supported by NAVF.

REFERENCES
1. Smith, M. (1985) *In vitro* mutagenesis. Ann. Rev. Genet. 19: 423-462.

2. Smith, M. (1986) Site-directed mutagenesis. Phil. Trans. R. Soc. Lond. A 317: 295-304.

3. Knowles, J. R. (1987) Tinkering with Enzymes: What are we learning? Science 236: 1252-1258.

4. Su, T.-Z. and El-Gewely, M. R. (1988) A multisite-directed mutagenesis using T7 DNA polymerase: application for reconstructing a mammalian gene. Gene 69: 81-89.

5. Nordvåg, B. Y., Nilsen, I., Husby, G. and El-Gewely, M. R. (1990) Screening of cDNA-libraries and gene reconstruction by PCR. This Proceedings.

6. Peden, K. W. C. and Nathans, D. (1982) Local mutagenesis within deletion loops of DNA heteroduplexes. Proc. Natl. Acad. Sci. USA: 7214-17

7. Shortle, D. and Botestin, D. (1983) Directed mutagenesis with sodium bisulfite. Meth. Enzymol. 100, 457-468.Smith, M. 1986. Site-directed mutagenesis. Phil. Trans. R. Soc. Lond. A. 317: 295-304.

8. Meyers, R. M., Lerman, L. S., and Maniatis, T. (1985) A general method for saturation mutagenesis of cloned DNA fragments. Science 229: 242-247.

9. Busby, S., Irani, M., and de Crombrugghe, B. (1982) I solation of mutant promoters in the *Escherichia coli* galactose operon using local mutagenesis on cloned DNA fragments. J. Molec. Biol. 154: 197-209.

10. Kadonaga, J. Y., Knowles, J. R. (1985) A simple and efficient method for chemical mutagenesis of DNA. Nucl. Acids Res. 13: 1733-1745.

11. Botstein, D., and Shortle, D. (1985) Strategies and applications of in vitro mutagenesis. Science 229: 1193-1201.

168

12. Barany, F. (1985) Single stranded hexameric linkers: a system for the in-phase insertion mutagenesis and protein engineering. Gene 37: 111-123.

13. Barany F. (1988) Procedures for linker insertion mutagenesis and use of new kanamycin resistance cassettes. DNA and Protein Engineering Techniques 1: 29-44.

14. Liu, C.-C., Simonsen, C. C. and Levison, A. D. (1984) Initiation of translation at internal AUG codons in mammalian cells. Nature 291: 349-351.

15. Saiki, R. K., Scharf, S., Faloona, F., Mullis, K. B., Horn, G. T., Erlich, H. A. and Arnheim, N. (1985) Enzymatic amplification of ß-globin genomic sequence and restriction site analysis for diagnosis of sickle cell anemia. Science 230: 1350-1354.

16. Saiki, R. K., Gelfand, D. H., Stoffel, S., Scharf, S. J., Higushi, R., Horn, G. T., Mullis, K. B., and Erlich, H. A. (1988) Primer-directed enzymatic amplification of DNA with a thermostable DNA polymerase. Science 239: 487-491.

17. Higuchi, R., Krummel, B., and Saiki, R. K. (1988) A general method for in vitro preparation and specific mutagenesis of DNA fragments: study of protein and DNA interactions. Nuc. Acids Res. 16: 7351-7367.

18. Kadowaki, H., Kadowaki, T., Wodisford, F. E., and Taylor, S. I. (1989) Use of Polymerase chain reaction catalyzed by *Taq* DNA polymerase for site-specific mutagenesis. Gene 76: 161-166.

19. Kamman, M., Laufs, J., Schell, J. and Gronenborn, B. (1989) Rapid insertional mutagenesis of DNA by polymerase chain reaction (PCR). Nuc. Acids Res. 17: 5404.

20. Vallette, F., Mege, E., Reiss, A. and Adesnik, M. (1989) Construction of mutant and chimeric genes using the polymerase chain reaction. Nuc. Acids Res. 17: 723-733.

21. Hagen-Mann, K. (1990) Application of the Polymerase chain reaction to DNA engineering. This Proceedings.

22. Hemsley, A., Arnheim, N., Toney, M. D., Cortopassi, G., and Galas, D. J. (1989) A simple method for site-directed mutagenesis using the polymerase chain reaction. Nuc. Acids Res. 17: 6545-6551.

23. Taylor, J. W., OTT, J., and Eckstein, F. (1985) The rapid generation of oligonucleotide-directed mutations at high frequency using phosphorothioate-modified DNA. Nucl. Acids Res. 13: 8765-8785.

24. Kunkel, T. A. (1985) Rapid an efficient site-specific mutagenesis without phenotypic selection. Proc. Natl. Acad. USA 82: 488-492.

25. Kunkel, T. A., Roberts, J. D. and Zakour, R. A. (1987) Rapid and efficient site-specific mutagenesis without phenotypic selection. Methods in Enzymol. 154: 367-382.

26. Olsen, D. B., and Eckstein, F. (1990) High Efficiency oligonucleotide-directed plasmid mutagenesis. Proc. Natl. Acad. Sci. USA. 87: 1451-1455.

27. Olsen, D. B., Sayers, J. R. (1990). Phosphorthioate-based mutagenesis for single- and double-stranded DNA vectors. This Proceedings.

28. Sanger, F., Niklen, S. and Coulson, A. (1977) DNA sequencing with chain-terminating inhibitors. Proc. Natl. Acad. Sci. USA 74: 5463-5467.

29. Tabor, S. and Richardson C. C. (1987) DNA sequence analysis with a modified bacteriophage T7 DNA polymerase. Proc. Natl. Acad. Sci. USA 84: 4767-4771.

30. Schultz, A. C. and Richards, J. H. (1986) Site-saturation of ß-lactamase: production and characterization of mutant ß-lactamases with all possible amino acid substitutions at residue 71. Proc. Nat. Acad. Sci. U.S. A. 83: 1588-1592.

31. Hill, E. D., Oliphant, A. R., and Struhl, K. (1987) Mutagenesis with oligonucleotides: An efficient method for saturating a defined DNA region with base pair substitutions. Methods in Enzymology 155: 558-568.

32. Anthony-Cahill, S.J., Griffith, M. C., Noren, C. J., Suich, D. J., and Schultz, P. G. (1989) Site-specific mutagenesis with unnatural amino acids. TIBS 14: 400-403.

33. Schold, M., Colombero, A., Reyes, A. A., and Wallace, B. (1984) Oligonucleotide-directed mutagenesis using plasmid DNA templates and two primers. DNA 3: 469-477.

© 1991 Elsevier Science Publishers B.V. (Biomedical Division)
Site-Directed Mutagenesis and Protein Engineering
M.R. El-Gewely, editor.

PHOSPHOROTHIOATE BASED MUTAGENESIS OF SINGLE AND DOUBLE-STRANDED DNA VECTORS

DAVID B. OLSEN & JON R. SAYERS

Max-Planck Institut für Experimentelle Medizin, Abteilung Chemie, Hermann-Rein Straße 3, D-3400 Göttingen, West Germany.

INTRODUCTION

The advent of automated oligonucleotide synthesis has made site-directed mutagenesis a routine technique. The molecular biologist may now alter any defined nucleotide sequence in a precise manner. The alteration, a mutation, may be a single transversion or transition change, multiple mismatch, or it may involve the insertion or deletion of one or more bases. The technique has applications in protein engineering, structure-function studies and in the study of molecular recognition. For example, alterations in promoter sequences, repressor binding sites and other regulatory sequences can be systematically introduced and the effect of the changes studied.

The basic principles of oligonucleotide-directed mutagenesis, as originally implemented by Zoller and Smith, may be summarized as follows (1). The DNA of interest is cloned into the double-stranded form of a single-stranded DNA bacteriophage such as one of the M13 series of phages(2). Recombinant single-stranded DNA is then packaged into phage and secreted into the growth media making isolation relatively simple. A mismatch oligonucleotide is used to introduce the desired change. The primer is complementary to the target sequence except for the mismatch position. This primer is then annealed to single-stranded recombinant DNA and the primer-template enzymatically converted to double-stranded closed circular DNA (RFIV). This heteroduplex RFIV DNA contains the desired mutant sequence on the viral (-) strand opposite the original wild-type sequence on the (+) strand. On transfection into a suitable cell line it is possible to obtain mutant progeny (1).

The percentage of phage progeny containing the mutation produced by this basic procedure is usually low. Most bacterial cell lines possess repair mechanisms capable of recognizing the wild-type methylated (+) strand of heteroduplex DNA and are able to reverse the mutation. The *in vitro* generated (-)stand carries no such methylated bases. This reduces the mutational efficiency to a level typically less than 10% (3).

Plaques generated by transfection of a heteroduplex specie may produce mixed genotype plaques due to the presence of wild-type and mutant sequences in the transfecting molecule. Such transformants require plaque purification prior to intense screening for positive clones. The phosphorothioate-based oligonucleotide-directed mutagenesis system which we describe avoids the limitations described above and can be used with either single or double-stranded DNA templates (4-9).

MUTAGENESIS OF SINGLE-STRANDED DNA VECTORS

Our mutagenesis method is based on the resistance of phosphorothioate internucleotidic linkages to enzymatic hydrolysis (10-14). Thus, double-stranded DNA containing phosphorothioate linkages in only one strand (introduced by *in vitro* polymerization) may be nicked in the non-substituted strand.

Scheme 1 outlines the procedure for the mutagenesis of single-stranded DNA vectors. In the first step a mismatch oligonucleotide primer is annealed to the DNA and is extended by an *in vitro* polymerization reaction in which one of the natural deoxynucleoside triphosphates is replaced by the corresponding deoxynucleoside 5'-O-(1-thiotriphosphate), dNTPαS. Thus, phosphorothioate groups are incorporated base specifically into one strand. The resultant strand asymmetry may then be exploited by a reaction with a restriction endonuclease (Table 1) which is unable to hydrolyze the phosphorothioate group at the position of cleavage in the (-)strand. The nicked (+)strand is then gapped with an exonuclease so that the wild-type sequence opposite the mismatch is removed. On repolymerization the gapped DNA is repaired using the mutated (-)strand as template. The fully complementary homoduplex DNA now contains the mutant sequence in both strands. Transfection with this DNA produces mutational frequencies of the order of 85% (5-8).

Several restriction endonucleases have been employed to produce a nick in the wild-type (+)strand (5-8,10-14). *Ava*I and *Nci*I have been used most frequently in this laboratory. Either exonuclease III (Exo III) or T7 gene 6 exonuclease (T7 exo) can be used for the gapping reaction. Amersham International supplies

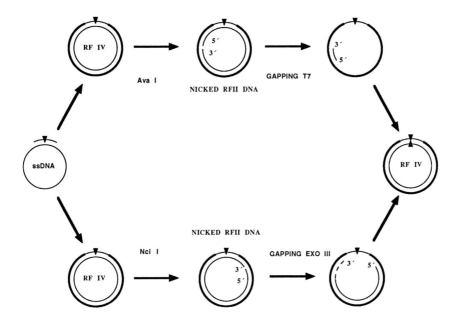

Scheme 1. Schematic representation of the phosphorothioate-based mutagenesis method for single-stranded DNA vectors. Single-stranded DNA, annealed with a mismatch primer, is converted to RFIV DNA using Klenow fragment, T4 DNA ligase and dCTPαS. The newly synthesized (−)strand is shown by the heavy lines. The (+)strand of the phosphorothioate-containing DNA is then nicked by reaction with a restriction endonuclease such as NciI or AvaI. The wild-type sequence is then digested away by either a 3′-5′ or a 5′-3 exonuclease. A fully complementary homoduplex RFIV molecule is generated on repolymerization.

a ready-to-use kit for the mutagenesis of single-stranded DNA which uses the enzyme combination NciI/Exo III. Exo III digests double-stranded DNA containing a 3' terminus in the 3'-5' direction at a rate of about 100 nucleotides per minute under the standard gapping conditions (15). Thus, a restriction site 1000 base pairs to the 3' side of the mutation within the (+)strand would require a reaction time of about 20 minutes (allowing a safety margin). T7 exo can also gap nicked double-stranded DNA, but in the opposite direction to exo III. It is capable of removing almost all the nicked (+)strand under normal conditions. Commercial samples of T7 exo appear to be endonuclease free and prolonged incubations are possible.

TABLE 1
List of Restriction endonucleases which are unable to linearize
phosphorothioate containing DNA.

Enzyme	DNA[a]	Polymerization Using	Reference[b]
AvaI	M13mp2	dCTPαS	11
	φX174	dTTPαS	11
AvaII	M13mp2	dGTPαS	11
EcoRV	M13mp18RV	dATPαS	14
FspI	M13mp18	dGTPαS	6
HindII	M13mp2,9,or 18	dGTPαS	6,11
NciI	M13mp2 M13mp18	dCTPαS	6,11
PstI	M13mp9,or 18 pUC19	dGTPαS	6,9,11
PvuI	M13mp2	dCTPαS	11
SmaI	M13mp18	dGTPαS[c]	8

a   The initial nicking conditions were determined using the DNA
vectors listed. Other restriction endonucleases such as; BamHI,
BanII, EcoRI, HindIII, PvuII and SacI, can also be us to nick
phosphorothioate containing DNA. See references 8, 12, 14.
b   It is recommended that all nicking reactions be carried out
according to the buffer and incubation conditions given in the
original reference.
c   Nicking using SmaI requires the presence of dGMPS at the sight
of cleavage and 40 µg/ml ethidium bromide in the reaction.

MUTAGENESIS USING PLASMID DNA VECTORS
    Creation of the mutant heteroduplex.
    The initial step in carrying out oligonucleotide-directed
mutagenesis of plasmid DNA vectors involves the site-specific
nicking of the DNA in the vicinity of the site of mutation. This
is carried out by using one of a number of restriction endonucleases which are unable to hydrolyze both strands of double-
stranded DNA when incubated in the presence of ethidium bromide
(16-20). Such reactions usually yield a small amount of linear
DNA with the main product being DNA which contains a nick in
either of the two strands (Scheme 2, B). Most of our experience
with this reaction has been with the enzymes EcoRI and HindIII
which are usually unique sites located at either the 3' or 5' end

of a number of multiple-cloning sites found in a variety of double-stranded vectors.

The nick created in the first step is then taken as the starting point for a gapping reaction with a nonprocessive 3'-5' exonuclease such as exo III or with the 5'-3' activity of T5 exonuclease (21). The extent of the gaping reaction catalyzed by such enzymes is regulated by the incubation time and after heat inactivation two distinctly gapped species are obtained, Scheme 2, C. The gapped specie which contains a small stretch of single-stranded DNA complementary to the ends of the mutant primer is considered "productively gapped". It is unimportant if the nick created in the first reaction is upstream or downstream from the site of mutation since the appropriate mutant oligo-nucleotide can be synthesized which is complementary to the productively gapped specie. It is advantageous to create the smallest gapped region possible so that mispriming of the mutant oligonucleotide to another single-stranded region of the DNA is minimized.

In vitro strand selectivity and formation of heteroduplex DNA

After strand-specific hybridization of the phosphorylated mutant oligonucleotide, mutant heteroduplex DNA is formed by ex-tension of the primer using the Klenow fragment and three dNTPs along with one dNTPαS analogue followed by ligation with T4 ligase. The choice of dNTPαS is governed by the choice of restriction endonuclease to be used in the next reaction. From this point the steps of the procedure are very similar to the procedure outlined earlier (see Scheme 1) for single-stranded DNA vectors. The introduction of phosphorothioate residues into the two distinctly gapped regions of the plasmid allows for *in vitro* strand-selective hydrolysis of all wild-type DNA in solution by reaction with a restriction endonuclease which is unable to cleave phosphorothioate internucleotidic linkages. This step linearizes roughly 50% of the DNA in solution. For the reaction depicted in Scheme 2 we have chosen *Pst*I, however, this reaction is not limited to the use of this enzyme (see Table 1).

Digestion of the nicked plasmid DNA using T7 exo removes the wild-type sequence which is not based paired with the mutant strand. Subsequent repolymerization with DNA polymerase I uses

176

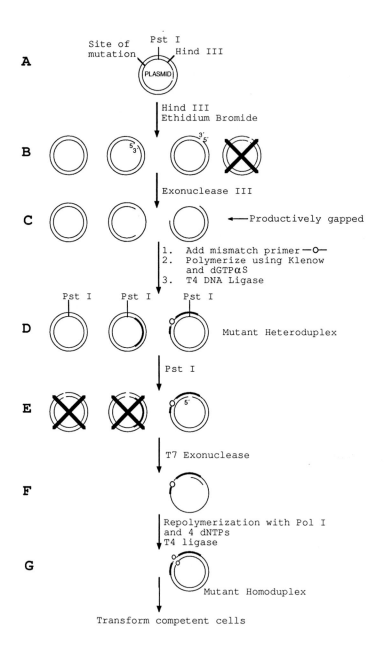

Scheme 2. Schematic representation of the oligonucleotide-directed plasmid mutagenesis technique. A, pUC19 double-stranded covalently closed circular DNA; B, products from HindIII reaction; C, products from exonuclease III reaction; D, mutant heteroduplex; E, products from PstI nicking/linearization reaction; F, products from T7 exonuclease reaction; G, repolymerized mutant homoduplex. The symbol "O" represents the mutation within the mismatch oligonucleotide. Heavy lines indicate the area where phosphorothioates have been incorporated. Taken from Ref.9.

the mutant strand as template yielding a mutant homoduplex ready for transformation, Scheme 2, G.

Experiments which consisted of several single and double-base substitutions into the polylinker or *lacZα* gene of the plasmid pUC19 have yielded efficiencies up to 70-80% (9).

Advantages of the plasmid mutagenesis procedure. Performing site-directed mutagenesis directly on plasmid DNA vectors has several advantages. Firstly, labor intensive subcloning steps into either M13 of the use of phagemid vectors can be avoided. In addition, the DNA remains in the double-stranded form during the annealing step. This decreases the likelihood of nonspecific hybridization of the mutant oligonucleotide. After polymerization the presence of dNMPS groups are localized to a small region of the DNA, resulting in good transformational efficiencies and high colony yields.

PRACTICAL CONSIDERATIONS

The preparation of high quality template DNA is an essential step in any site-directed mutagenesis experiment. Protocols for obtaining good quality single-stranded and plasmid DNA have been previously described (9,22). A high quality polymerase is required in the polymerization reaction. It should be free of all detectable 5′-3′ exonuclease activity. Such residual activity can digest the mutant primer and remove the mismatch. The Klenow fragment of *E. coli* DNA polymerase I, T4 DNA polymerase and T7 DNA polymerase (native enzyme) can all be employed.

MUTANT OLIGONUCLEOTIDE

The number of mismatches, or size of deletion or insertion is determined by the sequence of the mutant primer. For most cases an oligonucleotide with nine bases 3′-to the mismatch will protect it from the 3′-5′ exonuclease proofreading activity of most polymerases (15). To introduce single or double base mismatches we routinely use oligomers of 18-22 nucleotides with the mismatch positioned near the center. Larger deletions of up to 190 base pairs can be made using an oligomer of only 30 nucleotides (personal communication; T. Pope and M. Hetherington, Amersham International).

Another important concern is that the mutant oligonucleotide should not contain a high degree of self complementarity as this will cause problems in the annealing step. In addition, we suggest that at least two dG or dC residues be located (when possible) at the 5′ end of the oligonucleotide. This serves as a barrier to strand displacement by virtue of the higher melting temperature of G:C base pairs (23). Finally, the primer should not contain a recognition site for the restriction enzyme which is to be used for the hydrolysis of the wild-type DNA. Such a site would lead to the linearization of the DNA during the nicking reaction as the primer does not contain phosphorothioate groups.

SUMMARY

The phosphorothioate-based mutagenesis method has several advantages over other mutagenesis methods presently available. Firstly, the protocol can be used with either single or double-stranded DNA vectors. In addition, plaque purification is not needed and specialized cell lines are not required for transfection or for growth of the template DNA. Due to the high efficiency of mutagenesis it is usually sufficient to directly sequence 3 to 5 potential clones. Therefore, hybridization protocols or restriction analysis of large numbers of clones are not required.

ACKNOWLEDGEMENTS

We wish to express our thanks to Fritz Eckstein for directing the work described and to Wolfgang Pieken for critical reading of the manuscript.

REFERENCES

1. Zoller MJ, Smith, M (1982) Nucleic Acids Res. 10:6487-6500

2. Messing J (1983) Methods in Enzymology 101, pp 20-78

3. Friedberg EC (1985) DNA Repair. WH Freeman, New York, pp 614

4. Sayers JR, Eckstein F (1988) In "Genetic Engineering: Principles and Methods" Vol 10. Setlow JK (ed) Plenum Press, New York and London. pp 109-122

5. Taylor JW, Ott J, Eckstein F (1985) Nucleic Acids Res. 13:8765-8785

6. Nakamaye KL, Eckstein F (1986) Nucleic Acids Res. 14:9679-9698

7. Sayers JR, Schmidt W, Eckstein F. (1988) Nucleic Acids Res. 16:791-802

8. Sayers JR, Schmidt W, Wendler A, Eckstein F (1988) Nucleic Acids Res. 16:803-814

9. Olsen DB, Eckstein F (1990) Proc. Natl. Acad. Sci. USA 87:1451-1455

10. Potter BVL, Eckstein F (1984) J. Biol. Chem. 259:14243-14248

11. Taylor JW, Schmidt W, Cosstick R, Okruszek A, Eckstein F (1985) Nucleic Acids Res. 13:8749-8764

12. Olsen DB, Kotzorek GK, Sayers JR, Eckstein F (1990) J. Biol. Chem. in Press

13. Olsen DB, Kotzorek GK, Eckstein F (1990) Biochemistry, in press

14. Sayers JR, Olsen DB, Eckstein F (1989) Nucleic Acids Res. 16:9495

15. Gillam S, Smith M (1979) Gene 81-97

16. Dalbadie-McFarland G, Cohen LW, Riggs AD, Morin C, Itakura K, Richards JH (1982) Proc. Natl. Acad. Sci. USA 79:6409-6413

17. Parker RC, Watson RM, Vinograd J (1977) Proc. Natl. Acad. Sci. USA 74:851-855

18. Rawlins DR, Muzyczka N (1980) J. Virol. 36:611-616

19. Österlund M, Luthman S, Nilsson SV, Magnusson G (1982) Gene 20:121-125

20. Shortle D, Nathans D (1978) Proc. Natl. Acad. Sci. USA 75:2170-2174

21. Sayers JR, Eckstein F (1990) J. Biol. Chem., in press

22. Sayers JR, Eckstein F (1989) in Protein Function-A Practical Approach (Creighton, TE ed), IRL Press, Oxford. pp. 279-295

23. Fritz H-J (1985) DNA Cloning Vol I, Glover, D. M. (ed), IRL Press Limited, Oxford. pp 151-163

© 1991 Elsevier Science Publishers B.V. (Biomedical Division)
Site-Directed Mutagenesis and Protein Engineering
M.R. El-Gewely, editor.

MAPPING CATALYTICALLY IMPORTANT REGIONS IN ß-LACTAMASE USING TWO CODON INSERTION MUTAGENESIS

JOHN ZEBALA and FRANCIS BARANY

Department of Microbiology, Hearst Microbiology Research Center, Cornell University Medical College, 1300 York Avenue, New York, NY 10021 (U.S.A.)

ABSTRACT

Two-codon insertion mutants at eight positions in the ß-lactamase gene were characterized with respect to secretion and catalytic activity. Attempts were made to isolate second site revertants of various mutants. The activity of the mutants and the ability to obtain revertants showed a positive correlation with increasing distance from the active site based on the three dimensional structure of ß-lactamase. This observation is discussed as it may pertain to the generalized use of two-codon insertion mutagenesis in mapping important catalytic regions in enzymes.

INTRODUCTION

Understanding how protein structure gives rise to substrate specificity and catalysis is fundamental in the quest to engineer enzymes rationally. A basic tenet of protein structure is that the amino acid backbone and side chains pack tightly to form a highly stable structure (7). The amino acids in such a macromolecule may be thought of as providing both a spatial (structural spacing of the backbone) and compositional (specific sidechain interactions) contribution to the overall folding and activity of the enzyme. For example, catalytic amino acids would be expected to make a strictly compositional contribution to enzymatic activity, and substituting them with other amino acids would generally abolish activity (10,27). Structural amino acids very close to the active site would be expected to make mainly a spatial contribution by correctly positioning catalytic amino acids in the active site (11,25). As such, one would anticipate insertion mutations in these regions to be more deleterious than substitutions (26,28). Structural amino acids far from the active site may provide either a compositional or spatial contribution to folding and would probably be more tolerant of insertion mutations (2,28). Finally, amino acids in loops that serve as links between domains may have complete tolerance to insertions (2,3,28). Thus, we hypothesize that the activity of an insertion mutation will vary according

182

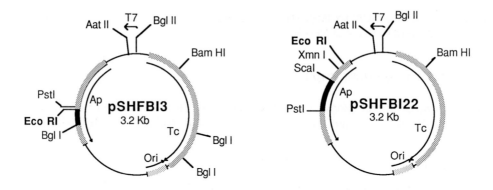

Fig. 1. Marker rescue of second site revertants in the ß-lactamase gene. Plasmids pSHFBI 3 and pSHFBI 22 are derivatives of pFBI3 and pFBI22, respectively (2) containing a T7 promoter upstream of the ß-lactamase gene (Ap). Each plasmid contained a two-codon insertion (EcoRI site) in the ß-lactamase gene which completely inhibited (pSHFBI 3) or severely reduced (pSHFBI 22) enzymatic activity. Plasmids containing ß-lactamase mutants were maintained in host cells by growth on tetracycline (Tc gene). Second site revertants of each mutant were selected as described in the text. Each second site revertant was localized within the gene (indicated as black region) by marker rescue of the Ap^R phenotype. Precise positions were determined by dideoxy sequencing as listed in Table 1 and Fig. 3.

TABLE I

TWO-CODON INSERTION MUTANTS AND REVERTANTS IN ß-LACTAMASE

PLASMID	LOCATION	TWO CODON INSERT				ACTIVITY	SECOND SITE REVERTANTS OF TWO CODON INSERT	ACTIVITY
pSHFB 2		WILD TYPE				++++		
pSHFBI 22	22	Ile	**GLU**	**PHE**	Glu	+	Met^{157}—→THR	+++
pSHFBI 5	53	Gly	**GLU**	**PHE**	Ala	-	—→ Loss of inserted two codons only	
pSHFBI 26	61	Ala	**GLU**	**PHE**	Gly	++		
pSHFBI 28	142	Pro	**ASN**	**SER**	Glu	+	—→ High plasmid copy only	
pSHFBI 3	165	Leu	**ARG**	**ILE**	Arg	-	Δ (Arg^{166}Lys^{167})—→Leu **ARG ILE** Leu	++++
pSHFBI 4	196	Leu	**ARG**	**ILE**	Arg	++		
pSHFBI 25	201	Pro	**ASN**	**SER**	Ala	++++		
pSHFBI 27	212	Ala	**GLU**	**PHE**	Gly	+	—→ [Ambiguous phenotype prevented selection of second site revertants]	

Location is amino acid prior to insertion. Amino acid numbering starts with the mature protein (e.g. Ile is the 22nd amino acid) based on the pBR322 sequence (32). Table is read from left to right in rows. For example, the Met(157) to Thr second-site revertant refers to pSHFBI22. No true second site revertants were isolated from either pSHFBI 5 or pSHFBI 28. Activities were determined as described (2). Amino acid changes for second site revertants were deduced from the DNA sequence determined by the dideoxy-sequencing method as described (35).

to its three dimensional distance from the active site, and may help identify active site and functional domains in a protein whose structure is unknown.

Several investigators have exploited site-specific mutagenesis to map important regions in proteins (for reviews, see 9,21,22,26,37). Although classical substitution mutagenesis may probe the compositional contribution of a particular amino acid, with the exception of active site residues, such a perturbation will generally be too subtle to effect activity. In contrast, two-codon insertion mutagenesis alters the spatial component of adjacent amino acids (1,2,4,5). For example, a two-codon insertion into an alpha helix will rotate subsequent amino acids by 205°, while an insertion into a ß-sheet simply extends the sheet. Despite the greater structural perturbation, in a wide variety of proteins studied, the majority of insertion mutants generated still contained some biological activity (2,3,5,6,8,12,15,18,28,29, 33,36,38). This indicates that while a two-codon insertion is structurally more severe than a substitution, it is not so severe as to abolish activity completely. It is this intermediate severity between point mutations and large deletions which makes a two-codon insertion useful as a probe to investigating structure-function relationships.

This work extends our studies on two-codon insertions in a small monomeric secreted protein, the pBR322 encoded ß-lactamase. This protein hydrolyzes penicillin and cephalosporin by transient acylation of its active Ser-45, as deduced from inhibitors (13), site-specific mutagenesis (10,11, 27), and the three dimensional structure (16). In addition, several investigators have isolated mutants that delineate the pathway of post-translational translocation into the periplasmic space (14,17,19,20,23,24, 30,31). Herein, we describe the characterization of two-codon insertion mutants in ß-lactamase, their second site revertants and the correlation of their enzymatic activity with the location of the active site from the three dimensional structure.

RESULTS

(a) Construction of plasmids for investigating ß-lactamase secretion. Two-codon insertions in the ß-lactamase gene gave a range of activities, from wild type to complete inhibition (Ref. 2 and Table 1). To investigate if aberrant protein secretion accounted for loss of activity, a T7 phage promoter was inserted upstream of several mutant genes (two examples are given in Fig.1). Use of a T7 phage promoter/polymerase system (34) allows one to shut down host transcription selectively with rifampicin, and thus exclusively label the ß-lactamase precursor and mature forms (Fig. 2). The

184

rate of ß-lactamase processing for a mutant containing a Glu-Phe insertion at position 61 (pSHFBI 5) was somewhat slower than a mutant containing the same amino acids at position 22 (pSHFBI 22) as well as wild type enzyme (pSHFB2). [Note that amino acid positions are numbered starting with the mature protein.]

    (b)   Isolation and characterization of second site revertants.
Revertants of ß-lactamase mutants with low or no activity were isolated by selecting for overnight growth at 32ºC or 37ºC on plates containing either 25µg/ml or 50µg/ml ampicillin. On average, about two to ten revertants were isolated per plate (10^9 cells plated). Since the original two-codon insertion creates an EcoRI restriction site, one could rapidly screen plasmids to check for primary site revertants (loss of the EcoRI site) or second site revertants (retention of the EcoRI site). Use of a RecA⁻ host (E. coli DH1 cells) severely reduced the number of primary site revertants isolated. Of the twenty to thirty revertants examined for each plasmid, the majority were primary revertants (loss of the original two codons), others gave high copy plasmids, and still others were chromosomal mutations. True second site revertants could only be isolated for mutants in pSHFBI 3 (two independent isolates) and pSHFBI 22 (three independent isolates). The location of these second site revertants was determined by ligation of specific restriction fragments into the primary mutant plasmid, and screening for rescue of ampicillin resistance (see Fig. 1). Dideoxy-sequencing of these regions revealed that both pSHFBI 3 revertants lost two adjacent codons, and that all three pSHFBI 22 revertants had a methionine to threonine change at position 157 (Table 1 and Fig. 3). The relative activities of mutants and revertants were determined by screening for growth on plates containing various concentrations of ampicillin, pipracillin, and keflin at 32ºC, 37ºC, and 42ºC as described (2).

DISCUSSION

    The activity of eight two-codon insertions in the ß-lactamase gene were characterized and second site revertants were selected, mapped and sequenced. Five of the insertions showed drastic reduction or loss of ß-lactamase activity as assayed on plates containing various ß-lactams (pSHFBI 28, pSHFBI 27, pSHFBI 5, pSHFBI 3 and pSHFBI 22). By examining the pulse/chase kinetics of processing the possibility of aberrant secretion was ruled out. These results are consistent with the interpretation that drug sensitivities of these two-codon insertion mutants are a direct consequence

pSHFB2 + pGP1-2   pSHFBI 5 + pGP1-2   pSHFBI 22 + pGP1-2

P   1   2   4   8   P   1   2   4   8   P   1   2   4   8

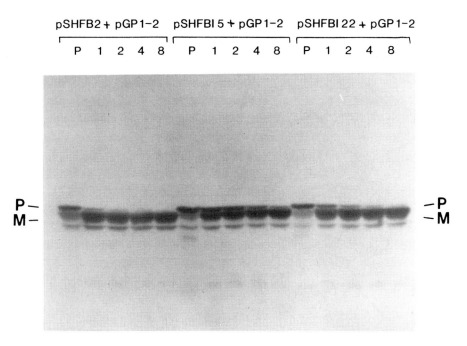

P—   —P
M—   —M

Fig. 2.   Autoradiogram showing kinetics of processing ß-lactamase mutants. E. coli cells containing a ß-lactamase under T7 promoter control (pSHFB2 = wild type) and an inducible T7 polymerase (pGP1-2) were grown in minimal media supplemented with 18 amino acids (0.1% without cys and met) at 32°C (34). After induction of T7 polymerase (42°C, 15 min), rifampicin was added to inhibit E. coli polymerase (200μg/ml, 42°C for 10 min, followed by 32°C for 30 min). Cells were pulse labeled with S^{35}-methionine (10μCi/ml cells) and chased with non-radioactive methionine. Samples were removed at various times during the chase, chilled to 0°C, harvested, resuspended in cracking buffer, and boiled for 3 min prior to loading on a 12% polyacrylamide gel containing 0.1% SDS. The gel was treated with Enhance (Du Pont), rinsed, and dried for autoradiography (Kodak XAR-5 film, 3 hr exposure). The pulse was for 30 sec. (P), and length of chase (min) indicated above each lane. Precursor [P] and mature [M] forms of ß-lactamase are indicated. The lower molecular weight bands (lighter) are unidentified, and have been observed previously (34).

of decreased enzymatic activity, and not with inefficient processing or secretion.

Of these five mutants, two (pSHFBI 28, pSHFBI 27) were close to the active site and one was on a secondary structural motif (helix 2) directly connected to the active site (pSHFBI 5) as judged from the three dimensional structure of ß-lactamase (16). True second site revertants could not be obtained for this class of two-codon insertions. In contrast, the other two mutants which gave low enzymatic activity (pSHFBI 22 and pSHFBI 3) were

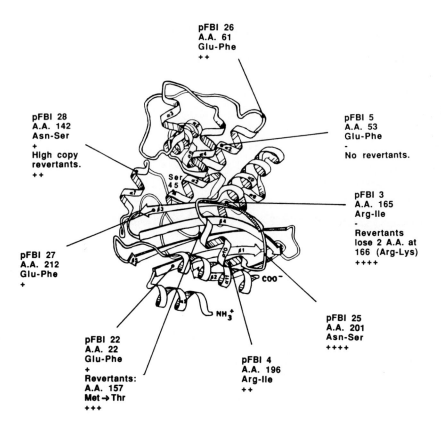

pFBI 26
A.A. 61
Glu-Phe
++

pFBI 28
A.A. 142
Asn-Ser
-
High copy
revertants.
++

pFBI 5
A.A. 53
Glu-Phe
-
No revertants.

pFBI 3
A.A. 165
Arg-Ile
-
Revertants
lose 2 A.A. at
166 (Arg-Lys)
++++

pFBI 27
A.A. 212
Glu-Phe
+

pFBI 25
A.A. 201
Asn-Ser
++++

pFBI 22
A.A. 22
Glu-Phe
+
Revertants:
A.A. 157
Met → Thr
+++

pFBI 4
A.A. 196
Arg-Ile
++

Ser
45

COO⁻

NH₃⁺

Fig. 3.   Position of insertion mutants and revertants on a ribbon diagram of ß-lactamase.
Drawing is from Herzberg and Moult (16) and used with the kind permission of J. Moult. Plasmid,
position of insertion, amino acids inserted, relative enzymatic activity, and nature of revertants
are indicated.

distant from the active site and revertants could be obtained for both (see
Table I). In the case of pSHFBI 3 the helix 8 is restored by the loss of two
codons adjacent to the original two-codon insertion. For pSHFBI 22 the
revertant is spatially close to the original two-codon insert. The three
insertions which gave good activity (pSHFBI 4, pSHFBI 25 and pSHFBI 26)
were both distant from the active site and between secondary structural
motifs.

These results are consistent with the conclusion that insertions close to the active site severely reduce enzymatic activity while those in between structural domains and distant from the active site gave good to wild type activity. Furthermore, true second site revertants could not be isolated from mutants close to the active site, but could be isolated for mutants far from the active site. This is not too surprising when one considers that the closer an insert is to the active site the less likely it is for a second site mutation to restore the active site geometry. The majority of two-codon insertions were sufficiently subtle to allow for residual enzymatic activity (an advantage when using a genetic screen). Thus, the finding of non-revertable two-codon insertions may be a general approach to mapping important catalytic domains in a protein whose structure in unknown.

ACKNOWLEDGEMENTS

We thank Soonjung Hahn and Lori Moran for work in the early parts of this study, Antje Koller for expert technical assistance and other laboratory members for helpful discussions. We thank Hamish Young for critical reading and suggestions. We thank Jeremy Knowles, John Moult, John Sondek, and Dave Shortle, for helpful discussion. This work is dedicated to the memory of E.T. Kaiser Jr. J.Z. is an MD-PhD student supported by the W.H. Keck Foundation. This work was supported by a grant from the National Institutes of Health (GM-41337-02) to F. B.

REFERENCES

1. Barany F (1985) Gene 37:111-123
2. Barany F (1985) Proc Natl Acad Sci USA 82:4202-4206
3. Barany F (1987) Gene 56:13-27
4. Barany F (1988) DNA and Protein Engineering Techniques 1:29-35
5. Boeke JD (1981) Mol Gen Genet 181:288-291
6. Boone C, Bussey H, Greene D, Thomas DY, Burnett P (1986) Cell 46:105-113
7. Clothia C (1975) Nature 254:304-308
8. Colicelli J, Lobel LI, Goff SP (1985) Mol Gen Genet 199:537-539
9. Craik CS (1985) Biotechniques 1:12-19
10. Dalbadic-McFarland G, Cohen LW, Riggs AD, Morin C, Itakura K, Richards JH(1982) Proc Natl Acad Sci USA 79:6409-6419
11. Dube DK, Loeb LA (1989) Biochemistry 28:5703-5707

12. Everett RD (1987) EMBO J 6:2069-2076
13. Fischer J, Belasco JG, Khoosla S, Knowles JR (1980) Biochemistry 19:4145-4149
14. Fitts R, Reuveny Z, Amsterdam JV, Mulholland J, Botstein D (1987) Proc Natl Acad Sci USA 84:8540-8543
15. Freimuth PI, Ginsberg HS (1986) Proc Natl Acad Sci USA 83:7816-7829
16. Herzberg O, Moult J (1987) Science 236:694-701
17. Kadonaga JT, Gautier AE, Straus DR, Charles AD, Edge MD, Knowles JR (1984) J Biol Chem 259:2149-2154
18. Kipreos ET, Lee GJ, Wang JYJ (1987) Proc Natl Acad Sci USA 84: 1345-1349
19. Koshland D, Botstein D (1982) Cell 30:893-902
20. Koshland D, Sauer RT, Botstein D (1982) Cell 30:903-914
21. Kramer W, Fritz H-J (1987) Methods in Enzymology 154:350-367
22. Kunkel TA, Roberts JD, Zakour RA (1987) Methods in Enzymology 154:367-382
23. Minsky A, Summers RG, Knowles JR (1986) Proc Natl Acad Sci USA 83:4180-4184
24. Pluckthun A, Knowles JR (1987) J Biol Chem 262:3951-3957
25. Schultz SC, Richards JH (1986) Proc Natl Acad Sci USA 83:1588-1592
26. Shortle D, DiMaio D, Nathans D (1981) Annu Rev Genet 15:265-294
27. Sigal IS, Harwood BG, Arentzen R (1982) Proc Natl Acad Sci USA 79:7157-7160
28. Sondek J, Shortle D (1990) Proteins: Struc, Func, and Genet 7:299-305
29. Stone JC, Atkinson T, Smith M, Pawson T (1984) Cell 37:549-558
30. Summers RG, Knowles JR (1989) J Biol Chem 264:20074-20081
31. Summers RG, Harris CR, Knowles JR (1989) J Biol Chem 264: 20082-20088
32. Sutcliffe JG (1979) Cold Spring Harbor Symp Quant Biol 43:77-90
33. Tanese N, Goff SP (1988) Proc Natl Acad Sci USA 85:1777-1781
34. Tabor S, Richardson CC (1985) Proc Natl Acad Sci USA 82:1074-1078
35. Tabor S, Richardson CC (1987) Proc Natl Acad Sci USA 84:4767-4771
36. Vieira J, Messing J (1982) Gene 19:259-268
37. Zoller MJ, Smith M (1987) Methods in Enzymology 154:329-350
38. Zumstein L, Wang JC (1986) J Mol Biol 191:333-340

© 1991 Elsevier Science Publishers B.V. (Biomedical Division)
Site-Directed Mutagenesis and Protein Engineering
M.R. El-Gewely, editor.

POLYMERASE CHAIN REACTION: A STRATEGY FOR SUCCESSFUL AMPLIFICATION

KERSTIN HAGEN-MANN

ESSC, Perkin-Elmer Cetus Analytical Instruments
Bahnhofstrasse 30, 8011 Vaterstetten, FRG

REPEATED CYCLES OF REPLICATION

The amplification of a target sequence via Polymerase Chain Reaction (PCR) involves synthesizing multiple copies of a region of DNA or a gene by a series of replication cycles, starting with oligonucleotide primers which bind to opposite strands that flank the target sequence. Each cycle of the reaction consists of denaturing the double-stranded DNA, annealing of the primers to their complementary target sequence on the single-stranded DNA, and extending the primers across the templates via a DNA polymerase. The newly synthesized DNA segments of which the terminus is defined by the 5'-end of the primers serve as templates for the next cycle. The result is an exponential amplification of the original target sequence. In 20-30 cylces it becomes possible to amplify the original sequence million-fold.

STANDARD PROTOCOL

Almost each amplification system needs to be adjusted for optimal results[1]. Therefore it is recommended to start with the standard protocol[2], analyze the PCR products, and then try to find the individual, system-dependent optimal conditions. One parameter only should be changed per optimization step!
A typical PCR reaction is set up in a 0.5ml microcentrifuge tube the final reaction volume being 50-100µl (covered by an mineral oil overlay) and containing the following:

−Target:	10pg-10µg DNA (cloned, genomic)
−Primer:	0.1-1.0 µM each
−dNTPs:	20-200 µM each (A,T,G,C), pH 7
−KCl:	50 mM
−Tris/HCl:	10 mM, pH 8.3 at room temperature
−$MgCl_2$:	1.5 mM
−*Taq* Polymerase:	1-2.5 Units

Standard incubation conditions:

After an initial denaturation of double-stranded template for 5-10 min at 95°C cycling should be performed as follows for n cycles:

Denaturation at 94°C for 1min

Annealing   at 37°C for 1min

Extension   at 72°C for 1min

A final extension step for 5-10 min after the last cycle is recommended.

Using these conditions and the standard buffer components a sequence of 150 to 3000 bases in length should be amplified successfully[3]. A 10µl aliquot of the reaction mixture must be enough to detect the product as a distinct band in an ethidium-bromide stained agarose gel. If there appears the band of expected lengh but also a smear of unspecifically amplified DNA the next step is to increase the annealing temperature.

PCR-TEMPLATE

In case the starting template is cloned DNA or highly purified, 10pg of starting material is sufficient for a reasonable yield of PCR product. But not only "clean" DNA can be amplified by PCR, also genomic DNA, crude cell lysates, and even RNA (with an additional cDNA synthesis step) can serve as template.

The number of starting molecules influences the number of cycles necessary to obtain detectable PCR products:

Table 1. Number of cycles dependent on the number of starting molecules[4]

number of starting molecules	number of cycles necessary to obtain 10^{12} molecules
10^5	25
10^3	30
1	45

PCR PRIMERS

More than any other component in the reaction the primers determine the success of the PCR. Selection of efficient and specific primers still is somewhat empirical; there are some computer programs available to design primers for a given target sequence and find the optimal annealing conditions.

Sometimes a correctly defined and properly synthesized primer is less efficient than expected and often a slight re-positioning of a new primer along the target sequence will solve the problem.

Primers of 18 to 30 bases length, with a random base dis-tribution and a G+C content similar to that of the fragment am-plified should be selected. Internal secondary structures or stretches of polypurines and polypyrimidines have to be avoided.

3' overlaps between the two primers may promote formation of the primer-dimer artifact, a short duplex product of the two-fold primer length.

Mismatches between the primer and the template are tole-rated by *Taq* DNA polymerase, only CC, GA, and AG mismatches between the 3'-end of the primer and the template will not be extended[5]. Wobble bases should not be situated at the 3'-end of the primers.

If the template sequence is not fully known, it is poss-ible to use degenerated primers[6] (mixture of oligonucleotides varying in the base sequence, but not in the number of bases). Mixed primers with a high degeneracy have been successfully ex-tended but the specificity of amplification will be the better the lower the primers are degenerated.

CARRYOVER AND CONTAMINATION

During PCR millions of copies of a target DNA are pro-duced. After amplification the reaction mixture contains very high concentrations of DNA and therefore must be treated care-fully. For each amplification a suitable positive and negative control should be chosen to avoid tube to tube cross-contami-nation which might generate false positive controls.

The PCR "set-up area" should be physically separated from the "PCR product-handling" area. Aerosol contamination by open-ing the tube after amplification is a serious problem. A safety cabinet with a germicidal lamp is recommended for the set-up of a PCR reaction. Water and pipette tips should be autoclaved. Also autoclave the stock solutions for the buffer separately before mixing them. All the components of the PCR reaction should be kept frozen in aliquots.

The use of positive displacement pipettors and gloves for handling PCR components and PCR products will also help to avoid contamination. Other laboratory equipment as gel-electro-phoresis apparatus, razor blades, centrifuges, UV trans-illuminator, and fume hoods are sources of carryover.

MAGNESIUM ION

The $MgCl_2$ has effect on the specificity of an amplification because *Taq* DNA polymerase is dependent on free magnesium ion concentration. With 200µM each dNTP usually 1.5mM $MgCl_2$ is optimal, but in some cases it must be adjusted for good results. With 800µM nucleotides present in the reaction, titration between 1,5-3,5mM $MgCl_2$ determines the optimal concentration. Too high magnesium concentration results in the accumulation of nonspecific products, too low magnesium in a lack of any PCR product.

REFERENCES

1. Williams JF (1989) Optimization Strategies for the Polymerase Chain Reaction. BioTechniques 7:762-768

2. Innis MA, Gelfand DH (1990) PCR Protocols: A guide to methods and applications. pp 3-12

3. Saiki RK (1989)In: Polymerase Chain Reaction. Erlich HA, Current Communications in Molecular Biology, CSH Lab. press, pp 25-30

4. Haff LA (1990) Perkin-Elmer Cetus, personal communication

5. Kwok S, Kellog DE, McKinney N, Spasic D, Goda L, Levenson C, Sninsky JJ (1990) Effects of primer-template mismatches on the polymerase chain reaction: Human immunodeficiency virus type 1 model studies. Nucl Acids Res 18:999-1005

6. Girgis SI, Alevisaki M, Denny P, Ferrier GJ, Legon S (1988) Generation of DNA probes for peptides with highly degenerate codons using mixed primer PCR. Nucl Acids Res 16:10371

© 1991 Elsevier Science Publishers B.V. (Biomedical Division)
Site-Directed Mutagenesis and Protein Engineering
M.R. El-Gewely, editor.

SCREENING OF cDNA-LIBRARIES AND GENE RECONSTRUCTION BY PCR

NORDVÅG, B.Y.*, NILSEN, I.*, HUSBY, G.**, EL-GEWELY, M.R.*

*Institute of Medical Biology, Dep. of Biotechnology, University of Tromsø. **Dep. of Rheumatology, University Hospital of Tromsø, University of Tromsø, Norway.

ABSTRACT

A human adult liver cDNA plasmid library was screened using PCR to isolate the normal human TTR-cDNA. Simultaneous trimming of the cDNA in both 5'- and 3'- ends was designed, making it suitable for site-directed mutagenesis and heterologous gene expression studies. Two appropriate PCR-primers, at the 5'- and 3'-ends of the target gene, were constructed. The PCR-product was digested with SphI and XbaI, and the resulting fragments indicated successful amplification of modified TTR-cDNA. PstI-digested PCR-product was ligated into both M13mp19 and pUC18 vectors. Positive hybridization signals were obtained in Southern blots of the DNA with inserts. DNA-sequencing of the cDNA cloned in M13, using the dideoxy method, revealed the successful cloning of the trimmed TTR-cDNA.

INTRODUCTION

The most commonly used method for cDNA library screening is by colony/plaque hybridization on nylon/nitrocellulose filters (1). However, the results are highly dependent upon the specificity of the hybridization probe and the method is time consuming and expensive. The polymerase chain reaction (PCR) was developed for the amplification of specific sequences of DNA (2). We have utilized this reaction in the screening of a cDNA-library for cloning of the normal transthyretin (TTR)-cDNA gene. In addition, a trimming of the gene was designed, making it suitable for site-directed mutagenesis and heterologous gene expression studies. The TTR protein is related to hereditary amyloidosis, which is a heterogenous disease, related to several different point mutations in the TTR protein (3). The TTR-cDNA gene was previously cloned and described by others (4).

MATERIALS AND METHODS

A human adult liver cDNA-library was screened using the PCR-reaction. $0.1 - 1\mu g$ template DNA was used, prepared by a plasmid minipreparation, using alkaline lysis method of a culture of the E. coli containing the library in PstI site of plasmid vector pKT218 (5). PCR-primers contained nucleotides complementary to some bases of the vector, included the PstI site, and some of the bases in each end of the TTR-cDNA (Fig. 1). Annealing temperatures in the PCR-reaction were 45° and 54°C. The PCR-product was analyzed by the restriction enzymes SphI and XbaI. The PstI-digested

194

PCR-product was ligated into phage M13mp19 or plasmid pUC18 and transformed into respectively DHα₅F and RR1 competent cells. The recombinant phages and plasmids were selected by absence of ß-galactosidase activity. Vector DNA's were isolated and examined for presence of inserts by gel-electrophoresis after PstI-digestion. Subsequently Southern blots of the DNA were hybridized, using a 5'-end labelled 22 bp oligo (P3) complementary to the middle region of the TTR-cDNA as a probe. DNA-sequencing using the dideoxy method, was made on DNA of recombinant M13 from several colonies.

RESULTS

According to the experimental design, the PCR-product gave a band of 496 bp and the restriction fragments had the expected sizes. Different annealing temperatures did not have any significant

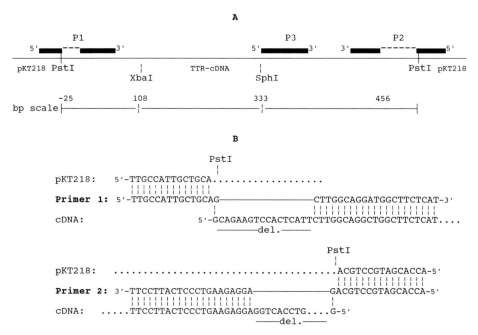

Figure 1 A: A diagrammatic presentation of TTR-cDNA and the flanking regions in vector pKT218. The positions of primers P1 and P2 for the PCR are indicated, and also that for the oligo P3 used in hybridization. Directions of the oligoes are indicated. Digestion points for restriction enzymes SphI and XbaI used are indicated in the diagramme. The restriction sites are shown on the approximate scale below. Truncated regions of P1 and P2 represent sequences of the cDNA that were trimmed out in the PCR strategy, resulting in the PCR product shown in Fig. 3.
B: Sequences of primers and parts of plasmid pKT218 vector and TTR-cDNA gene, demonstrating the design of primers regarding complementarity to vector and cDNA sequences and deleted parts of the cDNA. Primer 2 is complementary to the DNA-strand opposite of that shown for primer 1. 16 bp are deleted in the 5'-end by primer 1, while primer 2 deletes 133 bp in the 3'-end of the TTR-cDNA.

influence on the result. However, the highest annealing temperature gave a more distinct band (result not shown). Inserts of equal size, representing ligated cDNA, were observed as bands in the examined, PstI-digested recombinant plasmids and phages (Fig. 2). Strong hybridization signals were obtained in blots of insert-DNA from the examined plasmids and from four of seven phages (Fig. 2B). DNA-sequencing of the cDNA cloned in M13, revealed the sequence of the TTR-gene, as modified by the designed trimming (Fig. 3). A mutation of G to A_{76} was found in three out of four sequenced clones.

DISCUSSION

We present a new technique for screening cDNA-libraries by the PCR-reaction. We succeeded in cloning and simultaneous trimming of the TTR-cDNA from a human adult liver cDNA-library using PCR. The isolated, modified TTR-cDNA is suitable for further study by site-directed mutagenesis for heterologous gene expression.

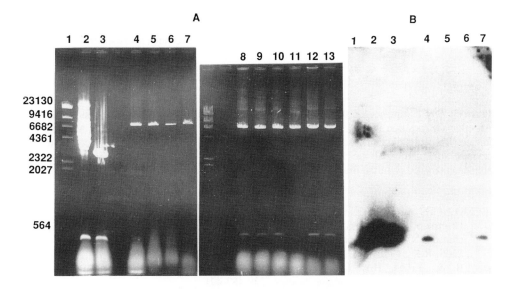

Figure 2: A. Agarose gel electrophoresis of DNA of pUC18 (lanes 2 and 3) and M13mp19 (lanes 4 to 13), examined for the presence of insert after subcloning and digestion with PstI. Lane 1 is a HindIII digest of λDNA. A significant band representing insert is visible in lanes 2-4,8-10, 12 and 13. (A band is also present in lane 7, but to weak to be visible in the picture.)
B. Southern blot of the DNA in A (lanes 1 -7), hybridized to the 5'-end-labelled hybridizastion-oligo (P3), complementary to the middle region of the cDNA. Strong signals were obtained from the inserts, and in addition from inserts in lanes 11 and 12 (not shown), indicating a successfull cloning of the trimmed TTR-cDNA.

Our developed technique (6) is a simple and quick tool in the screening of cDNA-libraries. This technique reduces the time used, lowers expenses and reduces the need for use of radioactive isotopes, compared with colony/plaque hybridization. This method could be utilized also in the cloning of unknown genes of interest, providing partial information about the protein sequence is availible to construct the primers.

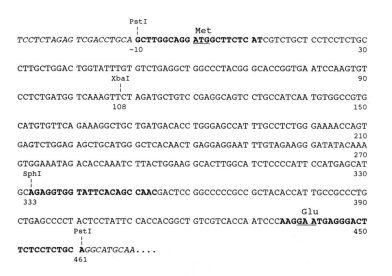

Figure 3: DNA sequence of the PCR-product after subcloning into M13mp19, using the PstI restriction site (position shown). The flanking regions of the vector are shown in italics. Met and Glu indicate first and last amino acids in the coding region of the cDNA-gene. Restriction sites of enzymes SphI and XbaI are indicated by |; bp numbers at the sites are given. Sequences complementary to, or in the primers/oligos, are indicated in bold types.

REFERENCES

1. Grunstein, M., Hogness, D.S. (1975) Colony hybridization: A method for the isolation of cloned DNAs that contain a specific gene. Proc. Natl. Acad. Sci. 72:3961.
2. Saiki, R.K., Gelfand, D.H., Stoffel, S., Scharf, S.J., Higuchi, R., Horn, G.T., Mullis, K. B. & Erlich, H.A. (1988) Primer-directed enzymatic amplification of DNA with a thermostable DNA polymerase. Science, 239:487-491.
3. Mita, S., Maeda, S., Shimada, K. & Araki, S. (1984) Cloning and sequence analysis of cDNA for human prealbumin. Biochem. Biophys. Res. Commun. 125:558-564
4. Costa, P.P., Falcâo de Freitas, A., Saraiva, M.J.M. (1990) Proceedings of the First International Symposium on Familial Amyloidosis. Arquivos Medicos, Porto, Portugal. Suppl. 3 in press.
5. Recombinant DNA Technical Bulletin, NIH (1978)
6. Nordvåg, B.Y., Nilsen, I, Husby, G., El-Gewely, M.R. (1990) Library screening and gene reconstruction by PCR. Manuscript in preparation.

© 1991 Elsevier Science Publishers B.V. (Biomedical Division)
Site-Directed Mutagenesis and Protein Engineering
M.R. El-Gewely, editor.

CHEMICAL MODIFICATION MAY BE USED TO STABILIZE AN ENZYME

Ciarán Ó'Fágáin & Richard O'Kennedy

School of Biological Sciences, Dublin City University,
DUBLIN 9, Ireland.

INTRODUCTION

The issue of continued biomolecule stability is vitally important
throughout Biotechnology. Stable enzymes and proteins capable of
maintaining acceptable performance over a long period are essential
for analytical devices and biosensors, for successful bioprocessing
and for reliable biological standards and controls.

Stabilized enzymes may be obtained by a variety of strategies, as
described in recent reviews [1,2,3]. Options narrow considerably,
however, if one wishes to use a stable enzyme in solution (rather than
immobilized) and when there is no suitable isolate available from a
thermophilic microorganism. Enzymes may be stabilized in solution by
judicious use of certain reagents [4] but these may lead to
undesirable side effects in some instances. Protein engineering by
site-directed mutagenesis has yielded stabilized enzyme derivatives
(see, for example, [5]). Protein engineering is, however,
time-consuming and generally relies on the availability of
considerable information concerning the target protein. Chemical
modification offers an alternative strategy for protein stabilization,
as pointed out by Mozhaev and colleagues [6].

Here we report on the enhanced thermal stability of chemically treated
derivatives of a clinically relevant enzyme, alanine aminotransferase
[L-alanine:2-oxoglutarate aminotransferase, E.C. 2.6.1.2, ALT, SGPT].
A report of this work has already appeared [7] and a full description
will be published shortly [8].

EXPERIMENTAL

Bis-imidates were obtained from Sigma. The crude preparation of beef
heart alanine aminotransferase was a gift from Baxter-Dade, Miami,
Florida, USA. Ion exchange celluloses were from Whatman UK and
Sephadex G-25 was obtained from Pharmacia. Enzfitter software, from
Biosoft-Elsevier, Cambridge, UK, was used for fitting of experimental
data.

Enzyme activity was measured spectrophotometrically at 340 nm using
the coupled kinetic assay of the International Federation of Clinical
Chemistry, with pyridoxal phosphate included in the reaction buffer as
an activator [9].

The method of Jenkins and Saier [10] formed the basis of the purification protocol. The crude enzyme preparation was desalted on Sephadex G-25 and then chromatographed on DEAE-cellulose in 20 mM Tris-HCl/10mM EDTA/20mM 2-mercaptoethanol, pH 7.0, using a 0 - 150 mM KCl gradient. ALT eluted soon after application of the gradient and the active pooled fractions were chromatographed on CM-cellulose in 60 mM acetate buffer/ 10 mM EDTA/20 mM 2-mercaptoethanol, pH 5.5, after their pH had been carefully lowered to 5.5 by the addition of HCl. ALT does not adsorb under these conditions and the pooled active fractions eluting from the column were concentrated using an Amicon YM-5 membrane following the return of their pH to 7.5. To assess purity, polyacrylamide gel electrophoresis was performed according to the Laemmli method [11].

The protocol for bis-imidates treatment was based on those of de Renobales and Welch [12] and of Minotani and colleagues [13]. ALT in 200 mM phosphate buffer, pH 8.0, was treated with bis-imidates in 400 fold molar excess for one hour at room temperature. Tris buffer, pH 8.0, was added after this time to ensure that the reaction had terminated. Accelerated storage tests to assess stability were set up and carried out according to the recommendations of Kirkwood [14] and were analyzed using the Enzfitter package. Enzyme kinetic determinations were carried out in triplicate over a 30 fold range of substrate concentrations and the Enzfitter programme was used to fit results to the Michaelis-Menten equation.

RESULTS

Recoveries of enzyme activity following bis-imidates treatment ranged from 88 - 96 %. Catalytic activities of the various enzyme fractions at elevated temperatures were expressed as a percentage of their 4 C-stored counterparts. Percent catalytic activity (relative to 4 C-stored fractions) was plotted against time and the data fitted well to a first order exponential decay curve in all cases. Table I shows the first order rate constants (per day) obtained from Enzfitter fits of experimental data.

TABLE I. FIRST ORDER RATE CONSTANTS FOR ALT ENZYME FRACTIONS
STORED AT DIFFERENT TEMPERATURES.

	No. $-CH_2-$ Groups	Length (Ångstroms)	First Order Rate Constant, k, at 45° C	37° C	27° C
CONTROL	-	-	.465	.125	.024
ADIPIMIDATE	4	7.7	.215	.069	.017
PIMELIMIDATE	5	9.2	.240	.065	.012
SUBERIMIDATE	6	11.0	.274	.066	.011

The kinetic constants Vmax and Km showed no significant alteration in value when compared with the untreated Controls. Vmax was found to be 44.9 ukat/l while Km values of 0.29 mM and 17.3 mM were determined for 2-oxoglutarate and L-alanine, respectively.

DISCUSSION

Beef heart alanine aminotransferase has been stabilized by treatment with bis-imidates. The basic catalytic parameters Vmax and Km have not been altered significantly. Native and modified ALTs follow a first order decay curve, consistent with a unimolecular inactivation mechanism.

Bis-imidates are known to react with protein amino groups with retention of positive charge [15]. It is likely that adjacently positioned amino groups within the enzyme backbone or between the subunits of this dimeric enzyme have been crosslinked by bisimidates treatment. Precise characterization of the stabilized derivatives is in progress. It would be instructive to investigate the effects of a wider range of chemical agents on ALT thermostability and to ascertain the molecular alterations in the stabilized derivatives. If individual crosslinked (or otherwise altered) residues were identified, a strategy for stabilization by site-directed mutagenesis could be devised.

Chemical modification is in many ways complementary to protein engineering, as recently discussed in a review describing the investigation of structure/activity relationships in a selection of proteins [16]. Many of the 20 amino acids can be modified quite specifically using mild and simple protocols with few or slow side reactions [17]. Our results suggest that chemical modification can play a complementary role to site-directed mutagenesis in the engineering of stabilized proteins.

Acknowledgment: We thank Baxter Dade, Switzerland and Eolas, the Irish science and technology agency, for financial support.

REFERENCES
1. Ó'Fágáin, C., Sheehan, H., O'Kennedy, R. & Kilty, C. (1988) Process Biochem. 23 166 - 171
2. Mozhaev, V.V. & Martinek, K. (1984) Enz. Microb. Technol.6 50 - 59
3. Torchilin, V.P. & Martinek, K. (1979) Enz. Microb. Technol. 1 74 - 82
4. Schein, C.H. (1990) Biotechnology 8 308 - 317
5. Oxender, D.L. & Fox, C.F.(eds.) 'Protein Engineering', Alan R. Liss, New York.
6. Mozhaev, V.V., Martinek, K. & Berezin, I.V. (1988) CRC Crit. Rev. Biochem. 23 235 - 284
7. Fernandez-Alvarez, J.-M., O'Fagain, C., O'Kennedy, R., Kilty, C. & Smyth, M.R. (1990) Anal. Chem. 62 1022 - 1026
8. O'Fagain, C., O'Kennedy, R. & Kilty, C. (1990) Enz. Microb. Technol., in press.

9. Bergmeyer, H.-U. & Horder, M. (1980) Clin. Chim. Acta 105 147F - 172F
10. Jenkins, W.T. & Saier, M. (1970) Meths. Enzymol. 17A 159 - 163
11. Laemmli, U.K. (1970) Nature 227 680 - 685
12. de Renobales, M. & Welch, W. (1980) J. Biol. Chem. 255 10460 - 10463
13. Minotani, N., Sekiguchi, T. Bautista, J.G. & Yokohama, N. (1979) Biochim. Biophys. Acta 581 334 - 341
14. Kirkwood, T.B.L. (1984) J. Biol. Stand. 12 215 - 224
15. Ji, T.H. (1983) Meths. Enzymol. 91 580 - 609
16. Profy, A.T. & Schimmel, P. (1988) Prog. Nucl. Acids Res. & Mol. Biol. 35 1 - 26
17. Imoto, T. & Yamada, H. (1989) in 'Protein Function:A Practical Approach', Creighton, T.E., ed., pp. 247 - 277

INDEX OF AUTHORS

SUBJECT INDEX